T0192176

Lecture Notes of the Institute for Computer Sciences, Social Informatics and Telecommunications Engineering 472

The LNICST series publishes ICST's conferences, symposia and workshops. It reports state-of-the-art results in areas related to the scope of the Institute.

LNICST reports state-of-the-art results in areas related to the scope of the Institute. The type of material published includes

- Proceedings (published in time for the respective event)
- Other edited monographs (such as project reports or invited volumes)

LNICST topics span the following areas:

- General Computer Science
- E-Economy
- E-Medicine
- Knowledge Management
- Multimedia
- Operations, Management and Policy
- Social Informatics
- Systems

Nishu Gupta · Prakash Pareek · M. J. C. S. Reis
Editors

Cognitive Computing and Cyber Physical Systems

Third EAI International Conference, IC4S 2022
Virtual Event, November 26–27, 2022
Proceedings

Editors
Nishu Gupta
Norwegian University of Science and Technology
Gjøvik, Norway

M. J. C. S. Reis
UTAD/IEETA
Vila Real, Portugal

Prakash Pareek
Vishnu Institute of Technology
Bhimavaram, India

ISSN 1867-8211 ISSN 1867-822X (electronic)
Lecture Notes of the Institute for Computer Sciences, Social Informatics and Telecommunications Engineering
ISBN 978-3-031-28974-3 ISBN 978-3-031-28975-0 (eBook)
https://doi.org/10.1007/978-3-031-28975-0

This Springer imprint is published by the registered company Springer Nature Switzerland AG
The registered company address is: Gewerbestrasse 11, 6330 Cham, Switzerland

Preface

We are delighted to introduce the proceedings of the third edition of the 2022 European Alliance for Innovation (EAI) International Conference on Cognitive Computing and Cyber Physical Systems (EAI IC4S 2022) hosted by Vishnu Institute of Technology, Bhimavaram, Andhra Pradesh, India during 26–27 November 2022. This conference brought together researchers, developers and practitioners from around the world who are leveraging and developing intelligent computing systems and cyber physical systems so that communication become smarter, quicker, less expensive, and accessible in bundles. The theme of EAI IC4S 2022 was "Cognitive computing approaches with data mining and machine learning techniques".

The technical program of EAI IC4S 2022 consisted of 22 full papers in oral presentation on web platform sessions at the main conference tracks. The above papers were presented by the registered authors in five technical sessions under five different tracks. The conference tracks were: Track 1 – Machine Learning and its Applications; Track 2 – Cyber Security and Networking; Track 3 – Image Processing; Track 4 – IoT Applications; and Track 5 – Smart City Eco-system and Communications. Apart from the high-quality technical paper presentations, the technical program also featured two keynote speeches and one invited talk. The two keynote speakers were Nishu Gupta from the Department of Electronic Systems, Faculty of Information Technology and Electrical Engineering, Norwegian University of Science and Technology (NTNU) in Gjøvik, Norway and Sara Paiva from the Polytechnic Institute of Viana do Castelo (IPVC), Portugal. The invited talk was presented by EAI IC4S 2022 Conference Chair Manuel J. Cabral S. from the Electrical Engineering Department of the University of Trás-os-Montes e Alto Douro (UTAD), Portugal.

Coordination with the steering chair, Imrich Chlamtac was essential for the success of the conference. We sincerely appreciate his constant support and guidance. It was also a great pleasure to work with such an excellent organizing committee team for their hard work in organizing and supporting the conference. In particular, the Organizing Committee chaired by Nishu Gupta, Prakash Pareek and Parvesh Kumar, the Technical Program Committee, chaired by Ahmad Hoirul Basori, and Ariel Soales Teres who completed the peer-review process of technical papers and made a high-quality technical program. We are also grateful to Conference Manager, Mikita Yelnitski for his support and all the authors who submitted their papers to the EAI IC4S 2022 conference.

We sincerely appreciate the management and administration of Vishnu Institute of Technology, Bhimavaram (VITB), Andhra Pradesh, India and especially the Chairman of the Sri Vishnu Educational Society (SVES) Shri K. V. Vishnu Raju, Vice Chairman of SVES, Sri Ravichandran Rajagopal, Secretary of SVES, Shri K. Aditya Vissam, and D. Suryanarayana, Director and Principal of VITB, for giving their support to us as host institute of EAI IC4S 2022.

We strongly believe that the EAI IC4S 2022 conference provided a good forum to all researchers, developers and practitioners to discuss all scientific and technological aspects that are relevant to cognitive computing and cyber physical systems. We also expect that the future EAI IC4S conferences will be as successful and stimulating, as indicated by the contributions presented in this volume.

March 2023

Nishu Gupta
Prakash Pareek
M. J. C. S. Reis

Organization

Steering Committee

Imrich Chlamtac	Bruno Kessler Professor, University of Trento, Italy

Organizing Committee

General Chair

Manuel J. Cabral S. Reis	UTAD University Engineering Department, Portugal

General Co-chair

Mohammad Derawi	Norwegian University of Science and Technology, Gjøvik, Norway

Organizing Chair

Nishu Gupta	Norwegian University of Science and Technology, Gjøvik, Norway

Organizing Secretary

Parvesh Kumar	Vaagdevi College of Engineering, Warangal, India

Technical Program Committee Chair

Ahmad Hoirul Basori	King Abdulaziz University, Rabigh, Saudi Arabia

Technical Program Committee Co-chairs

Prakash Pareek	Vishnu Institute of Technology, Bhimavaram, India
Ariel Soales Teres	Federal Institute of Maranhão, Brazil

Sponsorship and Exhibit Chair

Srinivasa Kiran Gottapu	University of North Texas, USA

Local Chair and Co-chairs

D. Suryanarayana	Vishnu Institute of Technology, Bhimavaram, India
S. Mahaboob Hussain	Vishnu Institute of Technology, Bhimavaram, India

Publications Chair

Raghvendra Pal	NIT Surat, India

Web Chairs

Ahmad Hoirul Basori	King Abdulaziz University, Rabigh, Saudi Arabia
Rakesh Nayak	O. P. Jindal University, Raigarh, India

Publicity and Social Media Chair

Pascal Lorenz	University of Haute Alsace, France

Workshop Chairs

Hendra Yufit Riskiawan	State Polytechnic of Jember, Indonesia
Raghvendra Pal	NIT Surat, India

Panels Chair

Parvesh Kumar	Vaagdevi College of Engineering, Warangal, India

Technical Program Committee

Neeraj Dhanraj Bokde	Aarhus University, Aarhus, Denmark
D. Suryanarayana	Vishnu Institute of Technology, Bhimavaram, India
Abhishek Kumar	Paris Saclay Université, Paris, France
Mukesh Sharma	Tel-Aviv University, Israel

Ankit Dixit	University of Glasgow, UK
Brojo Kishore Mishra	GIET University, India
Lakhindar Murmu	IIIT Naya Raipur, India
Nishit Malviya	IIIT Ranchi, India
Ashok Kumar	NIT Srinagar, India
Deepak Singh	NIT Raipur, India
N. Padmavathy	Vishnu Institute of Technology, Bhimavaram, India
D. J. Nagendra Kumar	Vishnu Institute of Technology, Bhimavaram, India
R. V. D. Ramarao	Vishnu Institute of Technology, Bhimavaram, India
R. Srinivasa Raju	Vishnu Institute of Technology, Bhimavaram, India
Venkata Naga Rani Bandaru	Vishnu Institute of Technology, Bhimavaram, India
G. K. Mohan Devarakonda	Vishnu Institute of Technology, Bhimavaram, India
Atul Kumar	IIITDM Jabalpur, India
Somen Bhattacharjee	IIIT Dharwad, India
Jitendra Kumar Mishra	IIIT Ranchi, India
Sumit Saha	NIT Rourkela, India
Pawan Kumar	NIT Rourkela, India
K. Srinivas	Vishnu Institute of Technology, Bhimavaram, India
Santos Kumar Das	NIT Rourkela, India
Amit Bage	NIT Hamirpur, India
Sridevi Bonthu	Vishnu Institute of Technology, Bhimavaram, India
P. Sita Rama Murty	Vishnu Institute of Technology, Bhimavaram, India
Idamakanti Kasireddy	Vishnu Institute of Technology, Bhimavaram, India
Pragaspathy S.	Vishnu Institute of Technology, Bhimavaram, India
Abhinav Kumar	IIIT Surat, India
Sadhana Kumari	CSIR National Aerospace Laboratories, Bengaluru, India
D. V. S. Bhagavanulu	Reva University, Bengaluru, India
Ashish Singh	Kalinga Institute of Industrial Technology (KIIT), Bhubaneswar, India
Vegesna S. M. Srinivasavarma	SRM Institute of Science and Technology, Chennai, India

B. V. V. Satyanarayana	Vishnu Institute of Technology, Bhimavaram, India
Siavanatha Maitrey	Vishnu Institute of Technology, Bhimavaram, India
G. Prasanna Kumar	Vishnu Institute of Technology, Bhimavaram, India
V. S. N. Narasimha Raju	Vishnu Institute of Technology, Bhimavaram, India
Sunil Saumya	IIIT Dharwad, Dharwad, India
Lokendra Singh	KL University, Vijaywada, India
Sunil Kumar	KL University, Vijaywada, India
A. Prabhakara Rao	Vishnu Institute of Technology, Bhimavaram, India
S. Sugumaran	Vishnu Institute of Technology, Bhimavaram, India
Yogesh Tripathi	KL University, Vijaywada, India
Abhishek Pahuja	KL University, Vijaywada, India
Ajay Kumar Kushwaha	Bharati Vidyapeeth, Pune, India
Amit Kumar	NIT Srinagar, India
Tanuja Satish Dhope	Bharati Vidyapeeth, Pune, India
D. M. Dhane	Bharati Vidyapeeth, Pune, India
Naveen Kumar Maurya	Vishnu Institute of Technology, Bhimavaram, India
N. P. Nethravathi	Reva University, Bengaluru, India
Sasidhar Babu S.	Reva University, Bengaluru, India
Gopal Krishna	Presidency University, Bengaluru, India
Priyanka Bharti	Reva University, Bengaluru, India
Laxmi B. Rannavare	Reva University, Bengaluru, India
Madhumita Mishra	Reva University, Bengaluru, India
D. R. Kumar Raja	Reva University, Bengaluru, India
Ashwin Kumar U. M.	Reva University, Bengaluru, India
Manoj Kumar Rana	SRMIST, Chennai, India
Veeramni S.	Amrita University, Chennai, India
Gireesh Gaurav Soni	SGSITS, Indore, India
Lalit Purohit	SGSITS, Indore, India
Megha Kuliha	SGSITS, Indore, India
R. C. Gurjar	SGSITS, Indore, India
Sumit Kumar Jindal	Vellore Institute of Technology, Vellore, India
P. Ramani	SRMIST, Ramapuram, India
T. J. Nagalakshmi	Saveetha University, Chennai, India
Rajkishor Kumar	Vellore Institute of Technology, Vellore, India
Avinash Chandra	Vellore Institute of Technology, Vellore, India
Naveen Mishra	Vellore Institute of Technology, Vellore, India

Ashish Mishra	KL University, Vijaywada, India
Abhishek Tripathi	Kalasalingam Academy of Research and Education, Krishnakoil, India
Vikram Palodiya	Sreenidhi Institute of Science and Technology, Hyderabad, India
Mayur Shukla	Oriental College of Technology, Bhopal, India

Contents

Image Processing

IoT Applications

Smart City Eco-System and Communications

Machine Learning and Its Applications

SQL Injection and Its Detection Using Machine Learning Algorithms and BERT

Srishti Lodha(✉) and Atharva Gundawar

Department of Computer Science and Engineering, Vellore Institute of Technology, Vellore 632014, Tamil Nadu, India
srishti2k1@gmail.com

Abstract. SQL Injection attacks target the database of applications to extract private information or inject malicious code. In this paper, we attempt to present a well-researched and practiced methodology to detect SQL Injection attacks accurately. These kinds of attacks are a very common means of network security attacks which can cause inestimable loss to the database. Building measures against them is a current research hotspot. Considering the possible complexity of queries involved and the need for a quick and efficient detection system in place, turning to machine learning techniques to combat and detect such attacks is the right choice. This is why we have undertaken the task of analyzing a number of machine and deep learning algorithms on a vast dataset of 41,770 points (consisting of both malicious and normal queries). We aim at finding a system that is accurate and fine-tuned for the best possible results and test each of the algorithms on various performance metrics to identify the one that performs the best. BERT outperforms the rest with a validation accuracy of 99.98%.

Keywords: Cyber-attack · Security · SQL Injection · Auto-detection · Machine Learning

1 Introduction

The volume of data used by Internet applications is increasing at a rapid rate. Time and resources are required to limit risks to privacy. Web applications typically keep their data in a centralized backend database, which makes them subject to many sorts of cyber assaults [16–18], specifically SQL injection (SQLi) attacks. One of the most hazardous injection attacks, SQLi undermines the fundamental security services of confidentiality, authentication, authorization, and integrity. In order to obtain access to or change data in a database, SQLi attacks include the insertion of harmful SQL instructions into input forms or queries. SQLIA are among the most dangerous vulnerability kinds, and pose significant risks to the security of online systems [22]. Several researchers have targeted this attack in their studies and discussed techniques for their detection and prevention [15, 19, 21, 27]. Multiple conventional security solutions, including firewalls, antivirus, anti-malware, etc. are also unable to detect this sort of attack sometimes, due to their

N. Gupta et al. (Eds.): IC4S 2022, LNICST 472, pp. 3–16, 2023.
https://doi.org/10.1007/978-3-031-28975-0_1

many levels of protection. SQLIA methods have grown more popular as it is simple to implement, and increasingly difficult to detect, demanding a need for an effective and practical solution in the domain of computer security. SQL injection attacks may be prevented or mitigated by ensuring that no fields are exposed to improper inputs and program execution. However, it is difficult to examine every page and application on a website manually, especially when updates are frequent and user-friendliness is important. Security experts and experienced engineers, recommend a number of procedures to guarantee that the database is adequately protected inside the bounds of the server. SQL Injection vulnerabilities in online applications may be avoided by using parameterized database queries with restricted, typed parameters and only using parameterized stored procedures in the database on rare occasions. This may be accomplished using a variety of programming languages, such as Java, .NET, PHP, and others.

1.1 Types of SQL Injection attacks

When attackers find a vulnerability in the system for the exploitation of SQL statements, they can launch SQLi in a number of ways [4, 20] (refer Table 1).

Table 1. Categories in SQLi attacks

Attack	Detail
Tautologies	The query always returns TRUE when the OR operator is used. Helps bypass user authentication and obtain data
Piggy-backed queries	Extra queries are added to the initial one, resulting in several SQL requests being sent to the database. The foremost query is legitimate and runs without any issues, but the following questions are the malicious ones running in parallel
Logically incorrect queries	Injecting unlawful or logically incorrect SQL syntax causes the program to display default error pages. This frequently exposes vulnerable parameters, thus helping attackers gather information to prepare for subsequent SQL injection attacks
Union query	In inserting a union query (statement injection attack), the attacker inserts an extra statement, say, of the kind "; <SQL statement>", into the original SQL line. Thus, a dataset which is a union of the original and injected queries is returned by the database
Stored procedure	Concentrates on the code of the database that can be exploited as computer code (or, stored procedure). The database engine executes stored operations directly. The stored method returns true or false, depending on whether the user is permitted or not. Post the first query, the illegitimate query forces the database to shut down

(continued)

Table 1. *(continued)*

Attack	Detail
Alternate encodings	Generates queries which make a database (or application) to respond in a different way, depending on the query's response, in order to obtain access to the database and find weak parameters. Two well-known approaches to do so are Blind-injection and Timing attacks

1.2 Fighting SQLi Attacks

To detect and prevent SQLI, there are many different techniques available [23]. Some of the most common ones are:

- Web framework – Special characters in an SQL query can be screened to recognize SQL injection threats.
- Static analysis – It works for tautology-based attacks, based on a dictionary-based detection method where an input query is flagged malicious as per fixed keywords. This technique has been enhanced over time with automated reasoning for better detection of SQL injection attacks.
- Dynamic analysis – It captures SQL queries from the client, database and the application, analyze the vulnerability, and SQL injection attack codes are used to comprehend the problem. The effectiveness of the attack is determined by comparing normal SQL queries to the attacked SQL queries. It is commonly used since it does not require any changes to the program and may operate independently, but the flaws must be addressed manually. Both static and dynamic analysis are used together to improve detection.
- Machine learning – SQL queries are learnt to create detection parameters. The produced SQL queries are then compared to the runtime SQL queries. Most SQLi attempts can be identified, depending on the strength of the provided parameters. There are other options, like employing a crawler-based machine learning technique [28].
- SQL profiling – It compares standardized SQL queries from a website to dynamic SQL queries runtime. Hence, there is no need to rewrite the web application, which if changes, will need the SQL query profile to be chained.
- Instruction-Set randomization – It inserts random numbers into the web application's SQL queries and evaluates the volatility, which can identify injection attempts. The proxy between the database and web server uses SQLrand. However, if the hacker predicts the random value, the approach is rendered worthless.

Apart from these, developers use a lot of tools installed on web and database servers [26], and evaluate SQL queries in real-time. These tools are regularly updated, and have a greater chance of detecting attacks than hard-coded approaches. Nevertheless, traditional methods are becoming obsolete, as skilled hackers constantly circumvent them and devise new methods of attack that can only be handled with automated and powerful security. This is why it is necessary to upgrade the defense mechanisms with Machine and Deep Learning technologies [24].

1.3 Organization of the Paper

The rest of the paper begins with the literature survey in Sect. 2, highlighting recent research work that employs ML in SQL Injection. The state-of-the-art is credited through a summarized table. In Sect. 3, we move to the proposed system and discuss the algorithms, including sub-sections about data acquisition and preprocessing, model tuning, etc. Next, in Sect. 4, we convey the results of this research, along with an analysis, and conclude the paper in Sect. 5.

2 Literature Review

Before proceeding to the proposed methodology of this work, it is important to review the existing studies similar to the current study. K. Kamtuo and C. Soomlek [1] used ML models on illegal queries, union queries, etc. on the server side. The 4 discussed models were susceptible SQLi commands on the server-side scripting. Nevertheless, building a compiler platform on IDE to validate and detect SQL syntax can further the scope of this study. B. A. Pham and V. H. Subburaj [2] proposed a methodology for detecting SQLi threats using 5 ML models, namely, Naive Bayes Classifier (NB), Support Vector Machines (SVM), Decision Forest, Logistic Regression (LR), and Extreme Gradient Boosting (XGB), and evaluated them using 5 metrics. However, in both these studies, the dataset was insufficient (<1500 data points). It is highly likely that the models are inefficient and overfitting occurred. Consequently, the performance might perform poorly when the model is used in the real world.

F. G. Deriba et al. [3] proposed a mechanism that combines static and dynamic detection strategies to validate a query. However, the paper lacks an in-depth discussion of this proposal, and the methodology followed during the implementation is not adequately discussed either. Although the proposed hybrid model is utilized to improve the overall model's efficiency, its training and testing time is also increased. I. Jemal et al. [5] is a review discussing several solutions proposed against SQL injection attacks, the type of attacks targeted, their mitigation measures etc. Nevertheless, the number of studies analyzed is not sufficient. Chen et al. [6] proposes a system for preventing SQLi attacks using NLP & Deep Learning techniques, aimed at reducing the false positives and providing some protection against new attacks. The Multilayer Perceptron (MLP) model performed better. Nonetheless, the researchers have not mentioned details about the model training. Moreover, very low number of studies have been referenced in this paper. A. Sivasangari et al. [7] develop the AdaBoost formula to examine various injection attack tactics. Although the study claims to have achieved better performance than other algorithms like KNN and NB, a lot of significant details like the dataset used, results, analysis, etc. are missing.

Xie et al. [8], implemented an Elastic-Pooling CNN-based model that can extract the most hidden common qualities of SQLi attacks which remain undetectable by the traditional models. Even though this model performed well, the training of this model was resource exhaustive. M. Hasan et al. [9], implemented an ML-based heuristic algorithm, trained on a small dataset. A GUI-based application is developed for 5 models. Ensemble Boosted Trees provided the highest accuracy (93.8%). More infected statements should be added in the dataset to improve the results. The authors of A. Falor et al. [10] explore various ML models compare them with their approach of using convolutional neural network (CNN) for SQLi detection. The CNN approach performed the best overall, with 94.84% accuracy, 85.67% precision, and 96.56% recall. Even though this model performs the best overall, none of these metrics in themselves are the best when compared to the rest of the ML models. Using bagging, we can combine the machine learning models to surpass all the performance metrics of the described approach. K. Zhang [11] reviewed ML techniques for the detection of SQLi attacks. It compares over 30 studies and 60 techniques from those papers, ignoring the datasets used in the target studies. Overall, an overview of the latest research is given and a table summarizes the ML algorithms, datasets, and evaluation techniques to compare the 11 methods.

Q. Li et al. [12] implemented an adaptive deep forest model or ADF to detect SQLi attacks. The authors only compare the model with simple DNNs, which aren't prepared to handle context over long inputs and this perform poorly than the ADF model. With the use of CNNs or RNNs, this gap can be overcome and neural networks will surpass the ADF approach in terms of all or most evaluation methods. A. Hadabi et al. [13] aims at preventing SQLIA with an adaptive model which uses runtime validation for the detection of these attacks. This model was only able to achieve an accuracy of 86.6%, on a dataset consisting of 4201 entries. ML mechanisms were not involved. We have overcome the limitations discussed in all these studies. The literature reviews are summarized below (Table 2).

Table 2. Summary of related works

Paper	Year of publication	Models employed	Highest validation accuracy	Most accurate model
K. Kamtuo and C. Soomlek [1]	2016	SVM, BDT, ANN, DT	99.68%	DT
B. A. Pham and V. H. Subburaj [2]	2020	NB, SVM, DT, LR, XGB	100%	DT, LR, XGB
F. G. Deriba et al. [3]	2022	NB, DT, SVM, ANN and Hybrid model	99.27%	Hybrid model
I. Jemal et al. [5]	2020	–	–	–
D. Chen et al. [6]	2021	MLP, CNN	98.57%	MLP
A. Sivasangari et al. [7]	2021	AdaBoost algorithm	–	AdaBoost algorithm
In X. Xie et al. [8]	2019	Elastic-Pooling CNN	99.93%	Elastic-pooling CNN

(continued)

Table 2. (*continued*)

Paper	Year of publication	Models employed	Highest validation accuracy	Most accurate model
M. Hasan et al. [9]	2019	Ensemble Boosted Trees, Cubic and Fine Gaussian SVM, Linear Discriminant Ensemble Bagged Trees,	93.8%	Ensemble boosted and bagged trees
A. Falor et al. [10]	2021	CNN, SVM, GNB, KNN, DT	94.84%	CNN
K. Zhang [11]	2019	DT, RF, SVM, LR, MLP, RNN, LSTM, CNN	95.4%	CNN
Q. Li et al. [12]	2019	Hybrid of ADF model and AdaBoost	97.5%	Proposed hybrid model
A. Hadabi et al. [13]	2022	Custom runtime model	86.6%	Custom model

3 Methodology

In this proposed system, we aim at optimizing ML and DL models, to classify incoming traffic as normal or malicious. The main advantage of using these techniques in the detection of the attack is the high accuracy and automated testing, thus reducing the requirement of human interference, while also maintaining significant security. Additionally, the use of a very large, unbiased dataset to train and validate our models helped in exposing the models to a wide variety of normal and malicious queries, thereby improving their performance in a real-time environment. The simple pipeline followed during the implementation is presented in the following figure (Fig. 1).

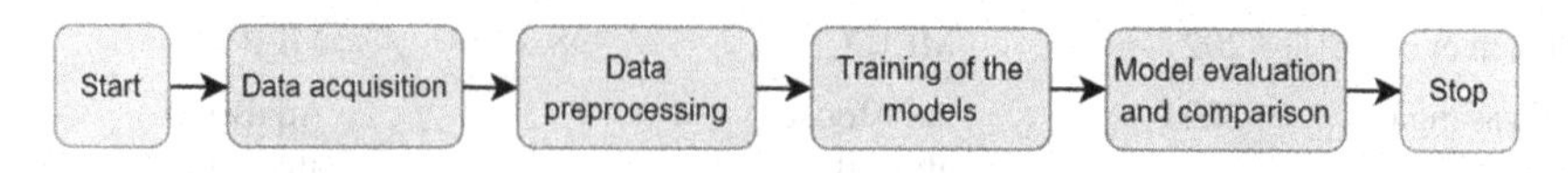

Fig. 1. Implementation pipeline

3.1 Data Acquisition

To increase the size of our dataset, we combined several datasets from different sources on Kaggle. We made sure that there was no mismatch of features and cleaned the datasets to bring them to a fixed format before merging. The final count of data points came up to 41,770, with 2 features: query and label. Figure 2 shows the distribution of malicious and non-malicious queries. The bias was kept low.

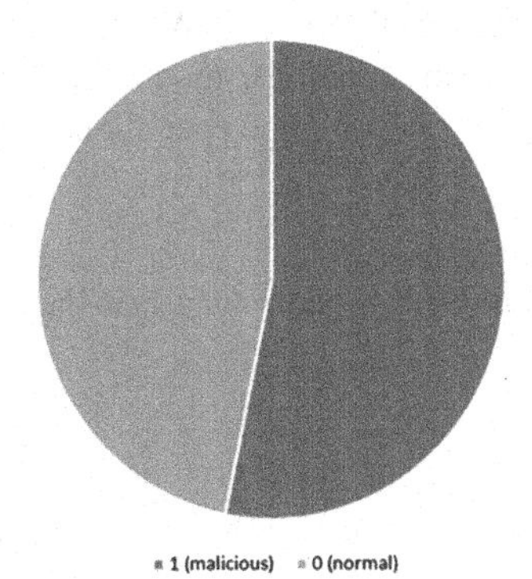

Fig. 2. Distribution of type of queries in the dataset

The malicious queries in the final dataset have all 5 types of SQLi attacks, categorized with their corresponding labels to identify their success or failure. The dataset was saved in.csv format. and imported with the Pandas package.

3.2 Data Preprocessing

As already discussed, the dataset that we have finalized for training our models has about 42,000 queries and has been conceived from the latest array of attacks so as to prepare our models for the current scenario. The CSV file created from these queries has been further cleaned to clear any missing values, and the unnecessary columns were dropped. Finally, exploratory data analysis was used to better interpret the data.

The queries are first run through a word tokenizer and split into word-level tokens according to the delimiter. This is necessary since almost all ML techniques only work on tokens. Now we can move on to encoding these using one hot encoding. This increases the efficiency of our codes and simplifies our whole process, since numbers are much easier to process than strings. The length of the longest question is obtained for the padding of the remaining queries.

Finally, once the data is split (test_size used: 25%, with shuffle 'true'), the shape of the training and validation datasets is confirmed and a very important step of data preprocessing is conducted: scaling. Using 'StandardScaler', the data is resized to obtain unit variance and zero mean. This was significant to standardize the values and drastically reduce the training time of the models. After completing these few steps under data preprocessing, we proceed to train our models.

3.3 Machine Learning Algorithms

Our analysis of the processed data is done using Scikit-Learn. Once we transferred our data into Scikit-Learn after following all the preprocessing steps mentioned above, we can classify it with the algorithms at our disposal. We have used SVM, Decision Tree (DT), Random Forest Classifier (RF), and K-Nearest Neighbours (KNN) for the binary classification with ML algorithms. BERT has been used as the DL classifier.

DT. DT is a decision support, tree-like model. We trained our data on an optimized version of the DT classifier. We fine-tuned the model using GridSearchCV, to find the best features from the parameter grid for some specific parameters (Table 3).

Table 3. Optimizing hyperparameters in the DT model

Parameter	Meaning	Default value
'max_depth'	Tree's maximum permitted depth	None
'min_samples_split'	Minimum samples needed before splitting any internal node	2
'criterion'	Function to measure split quality	'gini'
'max_features'	Helps determine the no. of features to be considered for best split	None

RF. While the DT algorithm generates only one tree, RF, an ensemble of such trees, generates multiple trees parallelly and takes into account the output of each of these trees to produce its own classification result. Table 4 shows the values tested from the parameter grid, as well as the search results for optimizing our RF model (using RandomizedSearchCV).

Table 4. Optimizing hyperparameters in the RF model

Parameter	Meaning	Default value	Parameter grid	Best value
'max_depth'	Maximum permitted depth of the trees	None	[5, 16, 27, 38, 50, None]	16
'min_samples_split'	Minimum samples needed to split an internal node	2	[2, 5, 10]	10
'n_estimators'	Number of trees	100	[100, 325, 550, 775, 1000]	100
'max_features'	Finds the no. of features to be taken for best split	auto	['auto', 'sqrt']	'auto'

(continued)

Table 4. (*continued*)

Parameter	Meaning	Default value	Parameter grid	Best value
'bootstrap'	Tells whether each tree is built using the whole dataset or bootstrap samples	True	[True, False]	False

KNN. KNN considers the classification of 'k' points from the dataset that are closest to a target point. Based on the majority of the class type observed, the target gets classified. Here, fine-tuning consists of choosing an optimal 'k' value to minimize the error rate and maximize the number of correctly classified points. We ran a loop to record the varying error rates with different 'k' values. A plot for the same can be observed in Fig. 3.

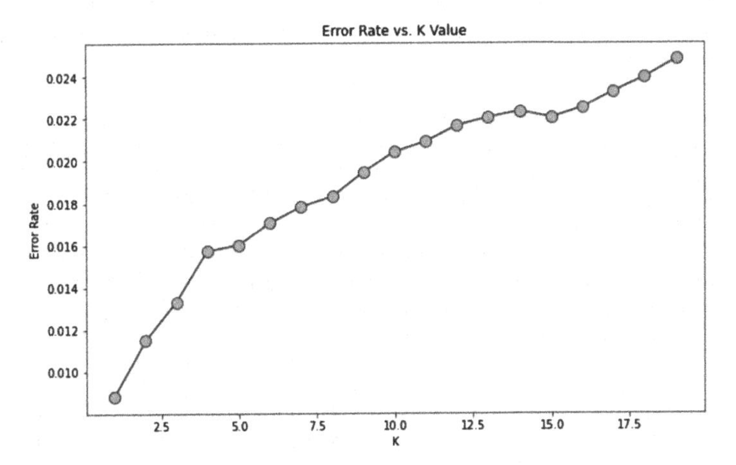

Fig. 3. Change in error with increasing K value for KNN algorithm

Clearly, we can see that the error increased consistently with an increase in the k values (which is not the case always). Hence, k = 1 is the best parameter for our situation, and we specify n_neighbors = 1 when training the model.

SVM. SVM is used for both type of problems, classification as well as regression. When used for classification, it differentiates between 2 classes by drawing a hyperplane on the dataset. Each observation (support vectors) helps in this computation. Due to the extremely high training time and computational requirement for SVM given the size of our dataset, we optimized the algorithm for just a few parameters (refer Table 5).

BERT. BERT (Bidirectional Encoder Representations of Transformers) [14] was developed and trained by researchers at google AI to tackle the problem of shortage of training

Table 5. Optimizing hyperparameters in the SVM model

Parameter	Meaning	Default value	Parameter grid	Best value
'C'	Penalty parameter of the error	1	[0.1, 1, 10, 100, 1000]	1000
'gamma'	Determines the impact of each support vector on the hyperplane	'scale'	['auto']	'auto'
'kernel'	Algorithm's kernel type to be used	'rbf'	['rbf', 'poly']	'rbf'

data in language models. BERT being the base model, converts text into a useful and concise context on which the classification head learns to make predictions. It is the first deeply bidirectional, unsupervised language representation model beats previous models at solving most language problems. BERT uses state-of-the-art attention models called transformers (that convert one input sequence into another) to map the contextual relations between words [25].

Transformers differentially weigh the significance of each part of the input using self-attention, thus retaining context over long input sequences and map the contextual relations between the multiple contexts. The architecture of transformers used in BERT is shown in Fig. 4. As explained, BERT can be used as a base model for a classification task. We can attach a classification head on the top of BERT and apply transfer learning techniques. We attach a word embedding layer to the pre-trained model before training the model, or add a custom function in the evaluation and training functions of the new model. We used the BertTokenizer for this research.

We began by creating a make model function defining the flow of data. Starting with the text input layer, we make a sequential model. Then we have a pretrained prepossessing layer and the BERT encoder layers. The context information is then passed on to the single-unit dense layer, which acts as the classification head. A sigmoid function is added to the final model as we are dealing with classification. BinaryCrossentropy Loss is used to compile the model, with the metrics being BinaryAccuracy. A custom Adam optimizer is made, where the training steps are equal to the data's cardinality. We set the initial learning rate to 4e−5, with 10% of the data's cardinality as the number of warmup steps. The model is compiled with the above-defined optimizer, loss function, and metrics. The training is done for 5 epochs only, which itself resulted in satisfactory accuracy. A simple threshold function was defined to convert values below 0.5 to 0 and others to 1 where 1 denotes SQL injection and 0 doesn't. An average of 60 ms were required for each training step and a total of 24,476 training steps were present in each epoch. The first epoch itself resulted in a validation accuracy of 99.67%, which only improved up to 99.80% till the 5th epoch.

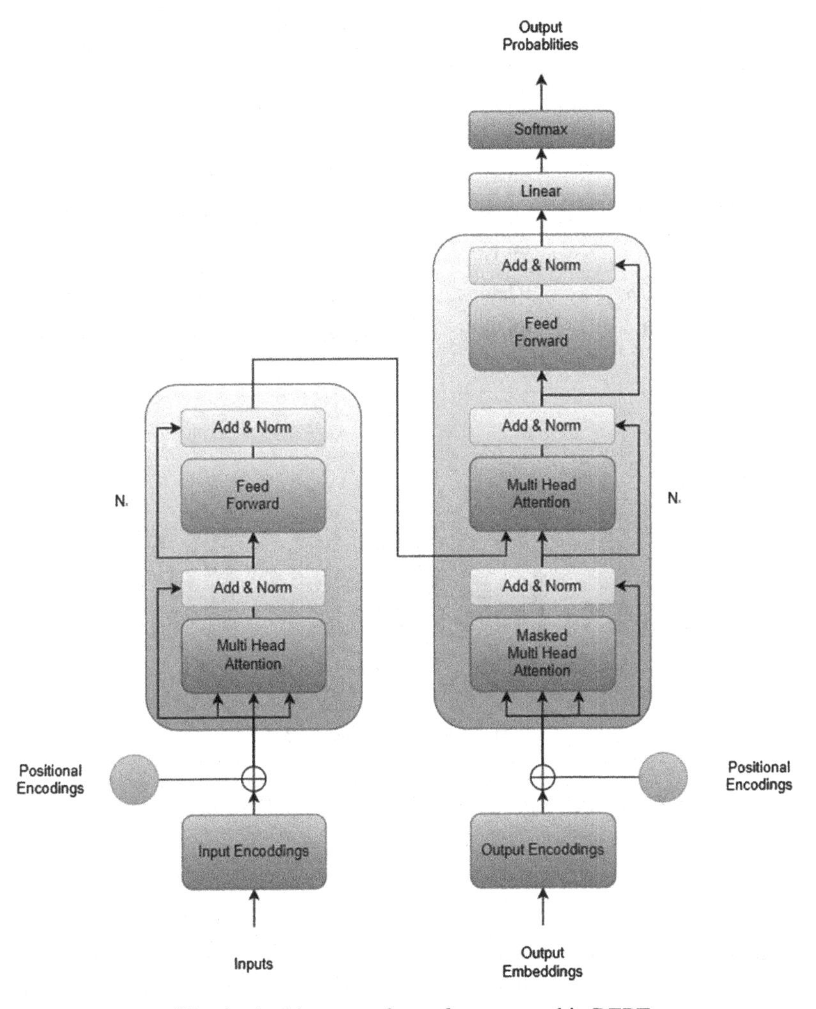

Fig. 4. Architecture of transformers used in BERT

4 Results

The research was completed using 5 algorithms (4 ML and 1 DL) for SQL detection. As already discussed, the dataset used was prepared by merging a number of datasets from Kaggle, which yielded a total of around 42,000 data points. The final dataset consisted of five types of malicious SQL queries. After rigorous training and hyperparameter tuning of the models, we observed that BERT performed the best on the processed dataset of the given problem. It was closely followed by the RF algorithm and the KNN classifier. The detailed performance metrics have been captured in Table 6 and visualized in Fig. 5.

Table 6. Comparing the model performances using various metrics (%)

Model name	Accuracy		Precision	Recall	F1
	Training	Validation			
DT	98.81	98.75	99.49	98.16	98.82
RF	99.93	99.79	99.93	99.68	99.80
KNN	100	99.12	98.85	99.52	99.18
SVM	92.78	92.44	90.21	96.36	93.19
BERT	**100**	**99.98**	**99.99**	**99.96**	**99.97**

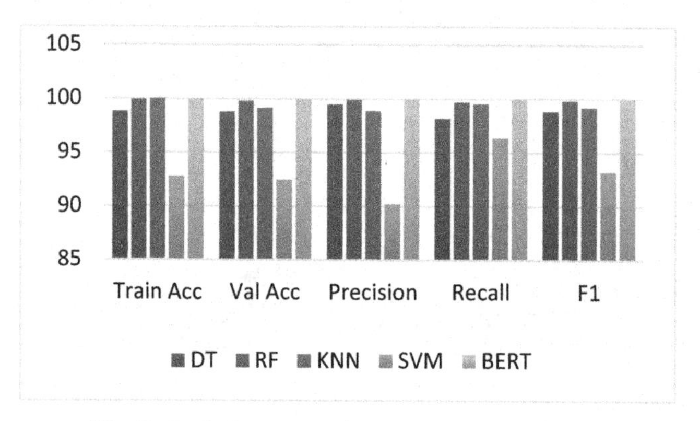

Fig. 5. Performance metrics achieved by the models

5 Conclusion

This research was conducted bearing in mind the threats that SQL injection attacks pose on our digital life. The best-automated method to detect SQL Injection attacks was discovered after running the acquired datasets through various machine and deep learning models. After an introduction about these attacks, their types as well as prevention mechanisms, a detailed literature survey of existing work was conducted. Post this, a simple methodology from data acquisition and data preprocessing to model training and validation was followed. Finally, the results were analyzed.

The BERT classifier surpassed every other model with a better accuracy, precision, recall, and the F1 score. Its validation accuracy (99.98%) was closely followed by Random Forest (99.179%). During the model training, we tackled problems regarding overfitting. We also eliminated the requirement of manual screening for feature extraction from our dataset. Nevertheless, this work has further scope to detect more attacks using the same model, but a much larger and diversified dataset. Additionally, more DL architectures can be explored for better performance and detection. Lastly, further research in this area can help enable these models to prevent SQLi attacks in addition to detecting them.

References

1. Kamtuo, K., Soomlek, C.: Machine learning for SQL injection prevention on server-side scripting. In: 2016 International Computer Science and Engineering Conference (ICSEC), pp. 1–6. IEEE (2016)
2. Pham, B.A., Subburaj, V.H.: An experimental setup for detecting SQLi attacks using machine learning algorithms. J. Colloquium Inf. Syst. Secur. Educ. **8**(1), 5 (2020)
3. Deriba, F.G., Kassa, T.M., Demilie, W.B.: Attacks on SQL Injection and Developing Compressive Framework Using a Hybrid and Machine Learning Approach (2022)
4. Hlaing, Z.C.S.S., Khaing, M.: A detection and prevention technique on SQL injection attacks. In: 2020 IEEE Conference on Computer Applications (ICCA), pp. 1–6. IEEE (2020)
5. Jemal, I., Cheikhrouhou, O., Hamam, H., Mahfoudhi, A.: SQL injection attack detection and prevention techniques using machine learning. Int. J. Appl. Eng. Res. **15**(6), 569–580 (2020)
6. Chen, D., Yan, Q., Wu, C., Zhao, J.: SQL injection attack detection and prevention techniques using deep learning. J. Phys. Conf. Ser. **1757**(1), 012055 (2021)
7. Sivasangari, A., Jyotsna, J., Pravalika, K.: SQL injection attack detection using machine learning algorithm. In: 2021 5th International Conference on Trends in Electronics and Informatics (ICOEI), pp. 1166–1169. IEEE (2021)
8. Xie, X., Ren, C., Fu, Y., Xu, J., Guo, J.: SQL injection detection for web applications based on elastic-pooling CNN. IEEE Access **7**, 151475–151481 (2019)
9. Hasan, M., Balbahaith, Z., Tarique, M.: Detection of SQL injection attacks: a machine learning approach. In: 2019 International Conference on Electrical and Computing Technologies and Applications (ICECTA), pp. 1–6. IEEE (2019)
10. Falor, A., Hirani, M., Vedant, H., Mehta, P., Krishnan, D.: A deep learning approach for detection of SQL injection attacks using convolutional neural networks. In: Proceedings of Data Analytics and Management, pp. 293–304. Springer, Singapore (2022)
11. Zhang, K.: A machine learning based approach to identify SQL injection vulnerabilities. In: 2019 34th IEEE/ACM International Conference on Automated Software Engineering (ASE), pp. 1286–1288. IEEE (2019)
12. Li, Q., Li, W., Wang, J., Cheng, M.: A SQL injection detection method based on adaptive deep forest. IEEE Access **7**, 145385–145394 (2019)
13. Hadabi, A., Elsamani, E., Abdallah, A., Elhabob, R.: An efficient model to detect and prevent SQL injection attack. J. Karary Univ. Eng. Sci. (2022)
14. Devlin, J., Chang, M.W., Lee, K., Toutanova, K.: Bert: pre-training of deep bidirectional transformers for language understanding. arXiv preprint arXiv:1810.04805 (2018)
15. Elshazly, K., Fouad, Y., Saleh, M., Sewisy, A.: A survey of SQL injection attack detection and prevention. J. Comput. Commun. (2014)
16. Jamil, A., Asif, K., Ashraf, R., Mehmood, S., Mustafa, G.A.: Comprehensive study of cyber attacks and counter measures for web systems. In: Proceedings of the 2nd International Conference on Future Networks and Distributed Systems, pp. 1–7 (2018)
17. Priyanka, A.K., Smruthi, S.S. Web application vulnerabilities: exploitation and prevention. In: 2020 Second International Conference on Inventive Research in Computing Applications (ICIRCA), pp. 729–734. IEEE (2020)
18. Kumar, S., Mahajan, R., Kumar, N., Khatri, S.K.: A study on web application security and detecting security vulnerabilities. In: 2017 6th International Conference on Reliability, Infocom Technologies and Optimization (Trends and Future Directions) (ICRITO), pp. 451–455. IEEE (2017)
19. Alwan, Z.S., Younis, M.F.: Detection and prevention of SQL injection attack: a survey. Int. J. Comput. Sci. Mob. Comput. **6**(8), 5–17 (2017)

20. Tasevski, I., Jakimoski, K.: Overview of SQL injection defense mechanisms. In: 2020 28th Telecommunications Forum (TELFOR), pp. 1–4. IEEE (2020)
21. Nasereddin, M., ALKhamaiseh, A., Qasaimeh, M., Al-Qassas, R.: A systematic review of detection and prevention techniques of SQL injection attacks. Inf. Secur. J. Glob. Perspect. 1–14 (2021)
22. Kareem, F.Q., et al.: SQL injection attacks prevention system technology. Asian J. Res. Comput. Sci. **6**(15), 13–32 (2021)
23. Marashdeh, Z., Suwais, K., Alia, M.: A survey on SQL injection attack: detection and challenges. In: 2021 International Conference on Information Technology (ICIT), pp. 957–962. IEEE (2021)
24. Alghawazi, M., Alghazzawi, D., Alarifi, S.: Detection of SQL injection attack using machine learning techniques: a systematic literature review. J. Cybersecur. Privacy **2**(4), 764–777 (2022)
25. Lin, J., Nogueira, R., Yates, A.: Pretrained transformers for text ranking: BERT and beyond. Synth. Lect. Hum. Lang. Technol. **14**(4), 1–325 (2021)
26. Baklizi, M., Atoum, I., Abdullah, N., Al-Wesabi, O.A., Otoom, A.A., Hasan, M.A.S.: A technical review of SQL injection tools and methods: a case study of SQLMap. Int. J. Intell. Syst. Appl. Eng. **10**(3), 75–85 (2022)
27. Oudah, M.A., Marhusin, M.F., Narzullaev, A.: SQL injection detection using machine learning with different TF-IDF feature extraction approaches. In: International Conference on Information Systems and Intelligent Applications, pp. 707–720. Springer, Cham (2023)
28. Urda, D., Basurto, N., Kull, M., Herrero, Á.: Evaluating classifiers' performance to detect attacks in website traffic. In: International Joint Conference 15th International Conference on Computational Intelligence in Security for Information Systems (CISIS 2022) 13th International Conference on EUropean Transnational Education (ICEUTE 2022): Proceedings, pp. 205–215. Springer, Cham (2023). https://doi.org/10.1007/978-3-031-18409-3_20

Solar Energy Prediction using Machine Learning with Support Vector Regression Algorithm

Idamakanti Kasireddy(✉), K. Padmini, R. V. D. Ramarao, B. Seshagiri, and B. Venkata Naga Rani

Vishnu Institute of Technology, Bhimavaram, India
kaasireddy.i@vishnu.edu.in

Abstract. Machine Learning is almost applied in every field such as engineering, science, medical etc. In this work, the concept of machine learning has been adopted for predicting solar energy. The solar Energy is widely known renewable energy due to its massive advantages. Solar energy prediction can help to determine the energy consumption beforehand and plays a major role in future planning. The grid operators are facing hardships because of unreliable weather conditions, which lead to the reduction in solar energy output. So, they are unable to satisfy the needs of consumers. Our proposed solution intends to make prediction models by using machine learning algorithms such as Linear Regression, Lasso Regression, Ridge Regression and Support Vector Regression (SVR). These algorithms use past weather data including temperature, dew, wind, cloud and visibility. Based on these data, analysis has been carried out in Jupyter Notebook. From the analysis, it has found that, SVR algorithm performed well when compared with other algorithms.

Keywords: Solar energy · Machine learning · Support Vector Regression (SVR)

1 Introduction

Solar energy has many advantages but it also has its own downside which is its production is highly irregular. Grid operators are facing problems because of the irregular output of solar energy as they are unable to meet the demand of consumers. But, by predicting solar energy grid operators can satisfy the consumers. If the production of solar energy at particular time is known then it is easy to plan according to the needs of users. There are many factors that affect the solar energy prediction. To predict solar energy one must collect data regarding these factors. The factors that are effecting the solar energy production involve wind, humidity, temperature and dew etc. Solar energy production can be predicted by analyzing the data which includes various factors that affect the solar energy with respect to the time and some other essential data. The data required can be collected from meteorological department or it is easily available in online. Machine

N. Gupta et al. (Eds.): IC4S 2022, LNICST 472, pp. 17–25, 2023.
https://doi.org/10.1007/978-3-031-28975-0_2

learning is the most effective approach to analyze data. It is classified into three types; they are supervised learning, unsupervised learning and reinforcement learning [1–6].

In supervised learning the model is provided with input data to get desired output. In unsupervised learning only input data is given to the model, leaving it on its own to find structure and in reinforcement learning a computer program interacts with a dynamic environment in which it must perform a certain goal. As it navigates its problem space, the program is provided feedback that's similar to rewards. In this study supervised learning has been used. Supervised learning is further classified into two types which include regression and classification. The most used regression algorithms are such as linear regression, lasso regression, ridge regression and support vector regression. Regression is basically predicting the value of dependent variables using independent variables.

Machine learning follows specific steps during its implementation, which includes data collecting, data analyzing, data wrangling, train & test and accuracy check. Firstly, the data is collected and collected data is analysed in order to check for duplicate values, Null values and wrong format. If null values are present in any rows of data those rows can be dropped or should be filled with appropriate values. The rows with duplicate values should be removed and wrong format is altered into correct format. This process is known as data cleaning. Then the model should be trained with 80% of data and remaining 20% is used to test the model. Finally the model undergoes accuracy check. Accuracy can be measured by using various methods such as R-Squared method. The algorithm is most efficient if it is more accurate.

Python simplifies the Machine Learning Algorithms by providing some libraries. Libraries that are used for machine learning are numpy, pandas, matplotlib, pyplot, sklearn and seaborn. These algorithms make machine learning algorithms easy to implement. Python is most used to implement machine learning algorithms due to its advantages such as python is easiest language and it provides many libraries. The platforms that are used to run machine learning algorithms include Anaconda, Google Colab and Jupyter. Among them Jupyter is simpler and it is very easy to share files using Jupyter Notebook. Jupyter is an open source which allows users to do mathematical computation such as trigonometry and Fourier transforms.

2 Methodology

2.1 Linear Regression

In this study we have used various regressions. Linear Regressions is one of the basic regression [7]. It helps to find the relation between independent and dependent variables. This regression is based on fitting best line into the graph and uses various methods to reduce error between best fit line and data points. Best fit line is also known as regression line. Linear Regression uses least-squares method to fit a line to the data and it uses R-squared value is the statistical measure of how close the data to the regression line. R-squared value is considered as accuracy and if its value is more than 0.5 than model is considered as good. It is used in Trend Forecasting, Evaluating trends and sales estimates.

Slope for the estimated regression equation is given by (1) and (2)

$$b_1 = \frac{\sum (x_1 - \overline{x})(y_1 - \overline{y})}{\sum (x_1 - \overline{x})^2} \tag{1}$$

$$\mathrm{R}^2 = 1 - \frac{SS_{RES}}{SS_{TOT}} = \frac{\sum_i (y_i - \underset{i}{\widehat{y}})^2}{\sum_i (y_i - \overline{y}_i)^2} \tag{2}$$

2.2 Lasso Regression

Lasso Regression [8] is used for Regularization. It is the technique to prevent model from getting over fitting. Sometimes the machine learning model works well with training data set but, when it is tested with testing data set it produces high cost function when compared to training data set hence it leads to over fitting. In case of Lasso regression the lambda is multiplying with the weights.

$$\sum\nolimits_{i=1}^{n} (y_i - \sum\nolimits_j x_{ij}\beta_{ij})^2 + \lambda \sum\nolimits_{j=1}^{p} |\beta_j| \tag{3}$$

2.3 Ridge Regression

Ridge Regression [8] is also used for Regularization. It is the technique to prevent model from getting over fitting. Sometimes the machine learning model works well with training data set but, when it is tested with testing data set it produces high cost function when compared to training data set hence it leads to over fitting. In case of Ridge Regression the cost function is changed by adding the penalty term to it. The amount of penalty added to the model is known as Ridge Regression Penalty. We can calculate it by multiplying with the lambda to squared weights

$$\sum\nolimits_{i=1}^{n} (y_i - \beta_0 - \sum\nolimits_{j=1}^{p} \beta_j x_{ij})^2 + \lambda \sum\nolimits_{j=1}^{p} \beta_j^2 = RSS + \lambda \sum\nolimits_{j=1}^{p} \beta_j^2 \tag{4}$$

2.4 Support Vector Regression

Support Vector Regression algorithms [9] are used to predict discrete values. Support Vector Regression uses the same principle as SVMs. In SVR we find the best fit line. Best fit line is the hyper plane that has the maximum no. of points. The SVR tries to fit the best line with a threshold value. The threshold value is the distance between hyper plane and boundary line. The advantages of SVR when compared to other algorithms are it is very easy to implement and it has high prediction capability.

2.5 Purpose

These four algorithms are analysed in this study to find out most accurate method for solar energy prediction. SVR turns out to be most accurate model among the other models.

3 Results

Firstly, in this analysis it is important to know about dependent and independent variables. In this data solar energy is the dependent variable and various factors affecting the solar energy are independent variables. Now, we will see the implementation of machine learning models.

Various regression algorithms are implemented and analysed for predicting solar energy in Jupyter Notebook. Initially, the python libraries such as numpy, pandas, matplotlib.pyplot and seaborn have been imported. The following commands are used to import libraries.

```
import pandas as pd
import numpy as np
import matplotlib.pyplot
import seaborn as sn
```

Then the solar weather data has been imported into the Jupyter notebook using the following command.

```
pd.read_excel('datainfo.xls')
```

Here the command varies with file type. If the file is csv type then the command changes to read_csv and name of the file should be written inside the brackets.

The imported data has been analysed using various commands such as info(), describe(), isnull().sum(). These methods are used to analyze data, data must be analysed in order to know about empty values, duplicate values and wrong formatted values. info() method is used to get information related to data such as type of value, is value null or non-null value, describe() is used to get information related to no of rows, max value, min value, standard deviation, 25%, 50% and 75% of values. The difference between values must be minimized in order to get more accuracy.

If any null values are present in data then the rows with null values must be removed or replaced with a suitable value. The rows with duplicate values must be dropped and rows with wrong format should be altered into correct format. In our data this type of values are absent.

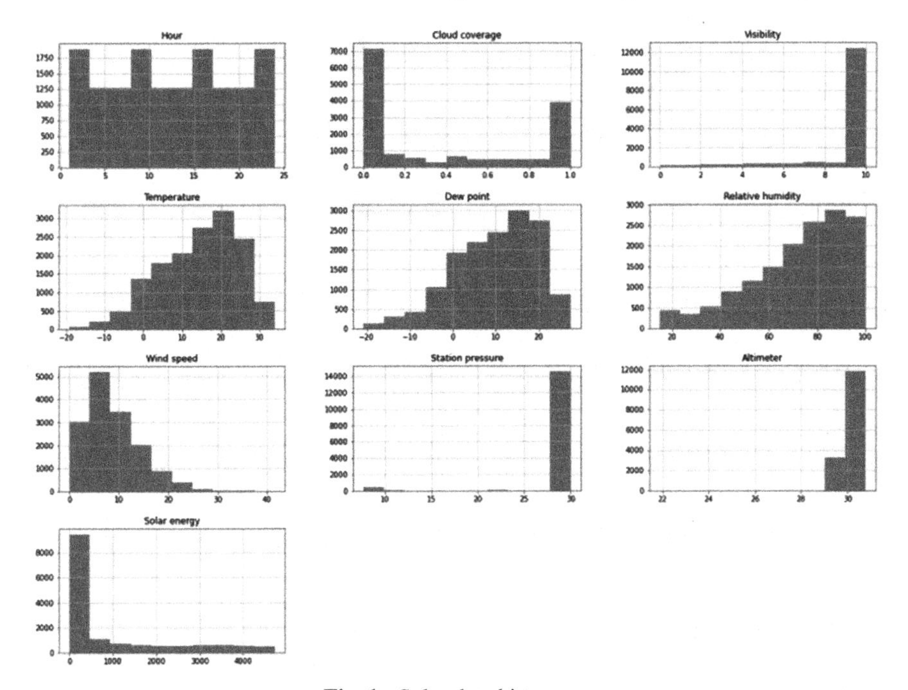

Fig. 1. Solar data histogram

The above graph shows the values present in columns of the data. The x-axis shows values in columns and y-axis shows no. of rows are present in data with same value. The above histogram can be drawn by using various methods present in matplotlib. pyplot library. Not only histograms but we can draw various graphs using the methods present in this library (Fig. 1).

Figure 2 shows the correlation between columns. If correlation between any columns is greater than 0.5 or -0.5 than they are highly correlated and also it varies from -1 to $+1$. Negative values stand for negative correlation and it shows inverse proportionality. Positive value indicates direct proportionality. The value of 1 is one to one relationship. Corr() method is used to find correlation and this method ignores the non-numeric values. In data non numeric values should be converted into numeric values for accurate analysis.

The training and testing of data can be done by importing train_test_split from sklearn.model selection library. It can be done by using following command.

$$from\ sklearn.model\_selectionimporttrain\_test\_split$$

The various algorithms can be directly implemented by using sklearn library. The following command is used to implement algorithms.

$$from\ sklearn.linear\_modelimportLinearRegression, Lasso, Ridge$$

These algorithms should be trained by using data.

$$lr = LinearRegression()$$
$$lr.fit(x\_train, y\_train)$$

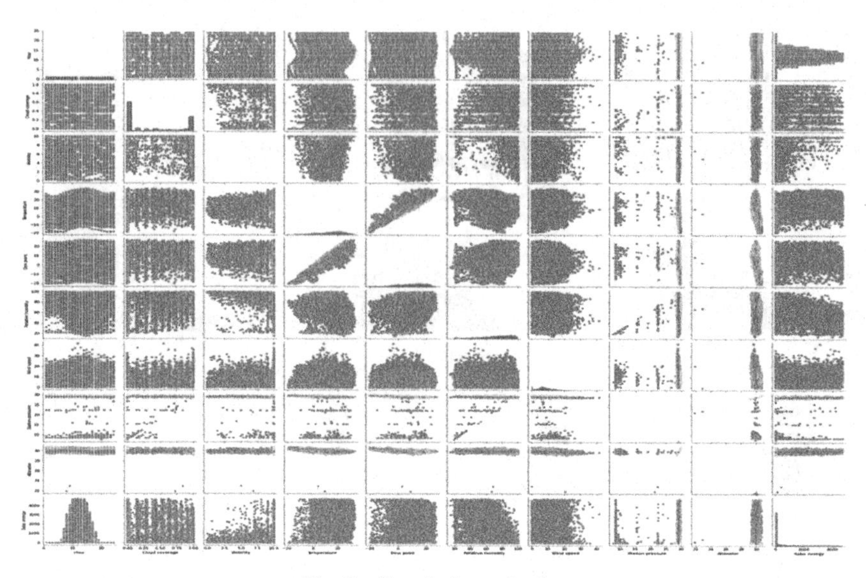

Fig. 2. Correlation pair plot

In the above command the linear regression is taken as lr and it is trained by using fit method. The same process follows for both Ridge and Lasso regressions.

$$\boldsymbol{las = Lasso(alpha = 0.1)}$$
$$\boldsymbol{las.fit(x\_train, y\_train)}$$
$$\boldsymbol{rid = Ridge()}$$
$$\boldsymbol{rid.fit(x\_train, y\_train)}$$

las represents lasso regression and rid represents ridge regression. As the training of model completed, next the testing of data should be done. For testing also various python provides various libraries.

$$\boldsymbol{las.predict(x\_train)}$$
$$\boldsymbol{las.score(x\_test, y\_test)}$$

In the above command the models is tested by using testing data and the accuracy of the model can be known by using above commands. The above commands are implemented for both Ridge and Linear Regression. SVR has to be imported from SVM as SVR is a part of SVM. For this also sklearn library is used and SVR also tested and trained in the same way as mentioned above.

Finally, the accuracies of various models are obtained as follows (Table 1).

Table 1. Comparison of various algorithms

Algorithm	Accuracy
Linear regression	51.3%
Lasso regression	51.2%
Ridge regression	51%
Support vector regression	88.4%

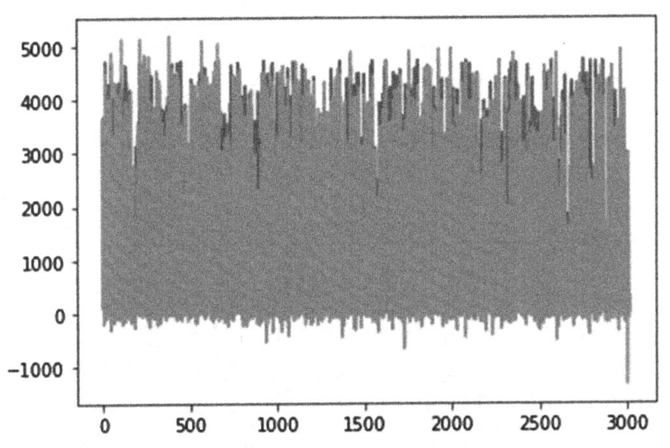

Fig. 3. Comparison of testing output (blue color) and predicted output (yellow color) for SVR

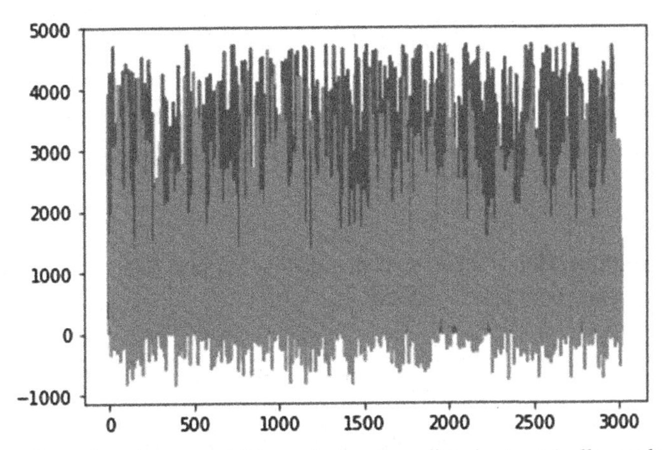

Fig. 4. Comparison of testing output (blue color) and predicted output (yellow color) for Linear Regression

Figures 3, 4, 5, and 6 depicts solar energy predictions for various algorithms. From Fig. 3, 4, 5 and 6, it is noticed that SVR algorithm outperforming when compared to

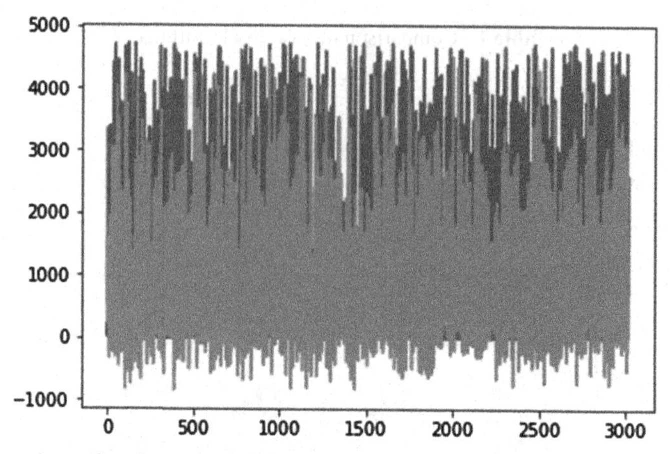

Fig. 5. Comparison of testing output (blue color) and predicted output (yellow color) for Lasso Regression

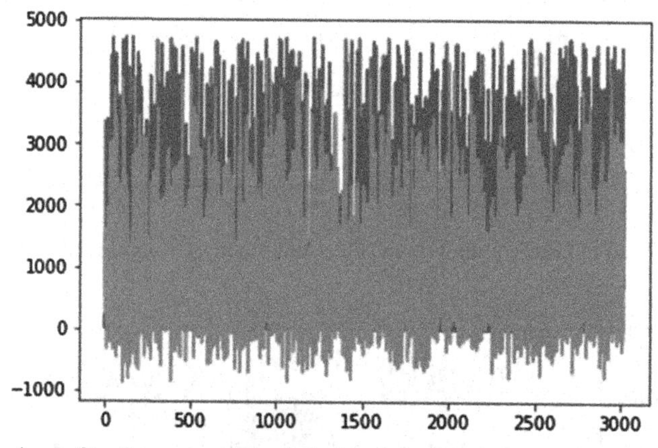

Fig. 6. Comparison of testing output (blue color) and predicted output (yellow color) for Ridge Regression

remaining algorithms. Hence it can be conclude that Support Vector Regression is most accurate model when compared to others.

4 Conclusion

In this work machine Learning models such as Lasso Regression, Ridge Regression, Linear Regression and Support Vector Regression are implemented and analyzed in Jupyter Notebook for solar energy prediction. From the analysis, accuracies of Lasso, Ridge and Linear regressions are found in between 51%–52%. Whereas the accuracy of Support Vector Regression is 88.4%. Hence it is clear that Support vector regression outperformed the remaining regressions models for predicting solar energy.

References

1. Javed, A., et al.: Predicting solar irradiance using machine learning techniques. In: 15th International Wireless Communications and Mobile Computing Conference (IWCMC), pp. 1458–1462 (2019)
2. Ak, R., et al.: Two machine learning approaches for short-term wind speed time-series prediction. IEEE Trans. Neural Netw. Learn. Syst. **27**(8), 1734–1747 (2016)
3. Nasir, A.W., Kasireddy, I., Rahul Tiwari, B.K., Ahmed, I., Furquan, A.: Data-based tuning of PI controller for first-order system. In: Bhaumik, S., Chattopadhyay, S., Chattopadhyay, T., Bhattacharya, S. (eds.) Proceedings of International Conference on Industrial Instrumentation and Control: ICI2C 2021, pp. 547–555. Springer Nature Singapore, Singapore (2022). https://doi.org/10.1007/978-981-16-7011-4_52
4. Sharma, A., Kakkar, A.: Forecasting daily global solar irradiance generation using machine learning. Renew. Sustain. Energy Rev. **82**(3), 2254–2269 (2018)
5. Kasireddy, I., et al.: Application of FOPID-FOF controller based on IMC theory for automatic generation control of power system. IETE J. Res. **68**(3), 2204–2219 (2022). https://doi.org/10.1080/03772063.2019.1694452
6. Kasireddy, I., et al.: Determination of stable zones of LFC for a power system considering communication delay. AIP Conf. Proc. **2418**, 040014 (2022). https://doi.org/10.1063/5.0081986
7. Maulud, D., Abdulazeez, A.M.: A review on linear regression comprehensive in machine learning. JASTT **1**(4), 140–147 (2020)
8. Yang, X., et al.: Lasso regression models for cross-version defect prediction. IEEE Trans. Reliab. **67**(3), 885–896 (2018)
9. Crone, S.F., Guajardo, J., Weber, R.: A study on the ability of support vector regression and neural networks to forecast basic time series patterns. In: Bramer, M. (ed.) IFIP AI 2006. IIFIP, vol. 217, pp. 149–158. Springer, Boston, MA (2006). https://doi.org/10.1007/978-0-387-34747-9_16

Price Estimation of Used Cars Using Machine Learning Algorithms

B. Valarmathi[1], N. Srinivasa Gupta[2], K. Santhi[3], T. Chellatamilan[3], A. Kavitha[4], Armaan Raahil[5], and N. Padmavathy[6](✉)

[1] Department of Software and Systems Engineering, School of Information Technology and Engineering, Vellore Institute of Technology, Vellore, Tamil Nadu, India
[2] Department of Manufacturing Engineering, School of Mechanical Engineering, Vellore Institute of Technology, Tamil Nadu Vellore, India
[3] Department of Analytics, School of Computer Science and Engineering, Vellore Institute of Technology, Vellore, Tamil Nadu, India
santhikrishnan@vit.ac.in
[4] Department of Mathematics, School of Advanced Sciences, Vellore Institute of Technology, Vellore, Tamil Nadu, India
kavitha@vit.ac.in
[5] Department of Information Technology, School of Information Technology and Engineering, Vellore Institute of Technology, Vellore, Tamil Nadu, India
[6] Department of Electronics and Communication Engineering, Vishnu Institute of Technology, Bhimavaram, Andhra Pradesh, India
padmavathy.n@vishnu.edu.in

Abstract. In this study, machine learning (ML) techniques are employed to predict used car prices. Several features are used to calculate the price of used cars, but in this paper, we find efficient ways to find the most precise car prices. Despite the fact that there are websites offering this service, they could not employ the most precise prediction system. It is also possible to predict a used car's true market value using a variety of models and techniques. It's important to understand their genuine market value before buying or selling. Both buyers and sellers will be benefitted from these accurate predictions. Support Vector regression, Random Forest regression, and CatBoost regression techniques are used in the proposed system. In the existing method [13], mean absolute error for decision tree regression was 0.6711, which was the least among other algorithms like Linear regression, Lasso regression, Ridge regression, Bayesian Ridge regression, and etc., they used. In the proposed system, mean absolute error (MAE) for Support Vector regression, CatBoost regression and Random Forest regression techniques are 0.1459, 0.1371 and 0.1284 respectively. The prices of second hand/used cars are predicted using the CatBoost regression, Support Vector regression, and Random Forest regression techniques. The accuracy of these algorithms are 86.28%, 85.40% and 87.16%. Among these three algorithms, Random Forest regression gives the least MAE of 0.1317 and the highest accuracy of 87.16%.

Keywords: CatBoost regression · Support Vector regression · Random Forest regression · Machine learning

N. Gupta et al. (Eds.): IC4S 2022, LNICST 472, pp. 26–41, 2023.
https://doi.org/10.1007/978-3-031-28975-0_3

1 Introduction

Many sites are involved in selling the used cars. CASH MY CAR is one of the largest car selling sites in India whose aim is to revolutionize the traditional method of selling used cars in India. On the basis of the different features that are used to determine the price, these companies have a mechanism for doing business with the buyer. As machine learning techniques have advanced, new algorithms have been developed that make it easier to make such predictions. Since the dataset has more features that might be used to estimate car prices, we want to use three machine learning techniques like Support Vector regression, CatBoost regression and Random Forest regression.

Yandex created the open-source boosting library is known as CatBoost, or categorical boosting. CatBoost can be utilized in ranking, recommendation systems, forecasting, and even personal assistants in addition to regression and classification. CatBoost is primarily concerned with decision tree theory and gradient boosting. The primary objective of boosting is to create a strong, competitive predictive model through greedy search by deliberately combining numerous weak models or models that just marginally outperform chance.

Support Vector Regression (SVR) is a supervised learning method for predicting discrete values. The same principle behind support vector machines underlies support vector regression. The core idea of SVR is to find the line that fits the data the best. The best fit line is generally regarded as the hyperplane with the highest number of SVR points.

A supervised learning technique known as Random Forest Regression uses the ensemble learning method for regression. In order to produce predictions that are more accurate than those from a single model, the ensemble learning method integrates predictions from several machine learning algorithms.

2 Literature Survey

In this paper [1], the main aim of the present research was to investigate various types of car data with the aim of developing an automated method for predicting car prices. They proposed a brand-new method for quantitative knowledge-based systems and data. They utilized a suggested car price model and provided non-numerical data solutions for forecasting. They used a range of data to solve the big data problem as well. Empirical predictions outperform other models in terms of performance.

In this research [2], they investigated how to forecast the cost of used cars in Mauritius using supervised machine learning techniques. The forecasts were created using historical information gathered from daily newspapers. Numerous techniques, such as K-nearest neighbours, Naive Bayes, and etc., were used for forecasting. The forecasts were then compared and assessed to see which ones delivered the best performances. The accuracy of handling a seemingly straightforward problem proved to be rather difficult. The performance produced by the four techniques was equal.

In this article [3], the need to estimate a used item's price according to its image and word-based description for different sets of item types, they provided a deep model architecture. For price prediction, this method used deep neural networks with long short-term memory (LSTM) and convolutional neural networks (CNN). In terms of mean absolute error accuracy, the model outperformed the standard support vector machine model. They suggested predicting the second-hand item's minimum and maximum price. Prediction task models use linear regression, seasonal autoregressive integrated autoregressive moving average and LSTM methods.

In this paper [4], they discussed the development, deployment, and assessment of ForeXGBoost. To enhance data quality, ForeXGBoost made full use of precisely crafted data filling algorithms for missing values. ForeXGBoost could increase forecast accuracy by extracting features from historical sales and production data using the sliding window. After an extensive investigation was undertaken to assess the impact of many variables on vehicle sales through the collection of information and the correlation of data, the most indicative features of the predictive set of features had been selected. Additionally, they quickly and accurately predict vehicle sales using the XGBoost prediction algorithm. Numerous tests showed that ForeXGBoost may produce accurate predictions with little overhead.

In this paper [5], Artificial Neural Networks, Support Vector Machines, and Random Forest are three machine learning approaches that were used to create an accurate model to estimate the prices of secondhand cars. Instead of using these approaches on individual data items, they were applied to the full collection of data items. Through a web site, which was also utilized to make the forecast, this data set was gathered. Web scrapers created in PHP programming language must be used to gather the data. To obtain the optimum outcome from the available data set, many machine learning methods with diverse capabilities had been compared. The last prediction model was added to a Java program.

In this paper [6], the authors explore how current deep learning models can be used to estimate a secondhand car's pricing as well as rate its performance and pollution index services. In the study, it was addressed how judgments about a car's performance can be made using an on-board diagnostic system and a variety of sensors. This would entail a general improvement in cost, emission, and performance predictions for automobiles.

In this article [7], using more than 100,000 data of used car sales from all around China; the authors carried out an empirical analysis and compared linear regression with random forest in detail. Three different models were employed with these two algorithms to predict the cost of used cars: one for a particular car series, one for a particular car make. In pricing evaluation models for particular auto manufacturers, the universal model, the findings showed that random forest clearly outperformed linear regression, but its effect is consistent but not perfect. As a result, it was demonstrated that the random forest method was the most effective for complex models with numerous variables and samples.

In this article [8], The proposed price evaluation approach was used to assess the pricing information for each type of car, which made use of widely dispersed car data and a substantial amount of vehicle transaction data. The process was intended to be developed in order to evaluate used car pricing and determine the price that best suited the car. The appropriate numbers of hidden neurons in the BP neural network (BPN) were selected using the optimized BPN approach to increase the precision of the prediction model and speed the convergence of the network topology. Through sample simulation trials, the actual transaction price generated from the improved model was compared with the fitting curve of the predicted price. Consequently, the optimized model's accuracy and fitting were both enhanced.

In this paper [9], they included feature screening and data preprocessing. Data transformation was used to standardize data formats in order to increase data quality. Data cleaning involved the removal of outliers and the filling in of missing values. The model building, training, and prediction processes all started with the screened features. Analysis of correlations and feature extraction using LightMBG were both part of the feature screening process. Five regression models were subsequently built, assessed, and the feature attributes collected through feature engineering for training. The result was a novel regression model that performed better than the five regression models when Random Forest and XGBoost were weighted and combined. Finally, the price of secondhand cars is predicted using the unique regression model.

In this paper [10], to forecast the cycle of trade in used cars, they applied machine learning. They combined the data, divided it into boxes, created additional features, dealt with outliers, and used principal component analysis to lower the dimension of the features in order to improve the expressivity of the model. Following that, the vehicle transaction cycle prediction data was fitted using the 50% discount cross-validation approach, and a random forest was employed for model training. It was discovered how each attribute related to the second-hand car transaction cycle. The cycle of transactions involving used cars was predicted and established. In both the test and the training sets, the fusion model's mean absolute error was 10.32 and 4.72, respectively.

In this paper [11], used car data was explored, and by examining the heat map, box line plot, and violin plot, features with correlations up to 99%, a few outliers, and missing values were discovered. Three distinct used car datasets were produced using three fill methods for missing values, namely, mode fill, median fill, and random forest fill, after the duplicate features and outliers were removed. Then, using the three used automobile datasets as input, four models—XGBoost, LightGBM, SVR, and BP neural network—were trained. The test set of used automobiles served as the subject of experiments, and the outcomes revealed that the LighGBM model performed well in both MAE and RA2, achieving results of 0.115350 and 0.965493, respectively. Which performed better on the LOSS of the training set than the other three models; yet, the BP neural network, perhaps as a result of the overfitting issue, on the test set performed poorly. The LightGBM improved the model on the XGBoost model to improve training speed and accuracy, so the results obtained by LightGBM model are slightly better than those of the LightBGM model. The SVR model performed best when the kernel selected was linear, but this did not apply to use car data, so the results obtained on MAE and R2 were worse than those of the other three models.

Developed a mathematical model that could estimate the cost of a used car based on its current attributes was the aim of this study [12]. It was challenging to estimate the cost of a used car because several variables, including current mileage, condition, make, and year, among others, might affect the price. Furthermore, precisely predicting the price of a used car became problematic from the standpoint of a seller. Therefore, the focus was on developing tools and researching models that could accurately predict a used car's price based on its capabilities. As a result, a customer could make a purchase with considerably more knowledge.

In this article [13], a large number of features and parameters needed to be taken into account in order to produce accurate results made it difficult to predict used car prices. The initial and most crucial phase was data gathering and preprocessing. After that, a model for formulating algorithms and generating output was created. Different regression techniques were used on the model, and the best performance was found for the Decision Tree Algorithm.

3 Dataset Description and Sample Data

The dataset includes nine attributes namely Car Name, Year, Selling Price, Present Price, Kms Driven, Fuel Type, Seller Type, Transmission and Owner. 301 rows are present in this dataset. The description of the nine attributes of the dataset is shown in the Table 1. In the dataset, the attributes like Car Name, Fuel Type, Seller Type and Transmission contains the categorical values. The sample dataset before preprocessing is shown in the Table 2. For example, let's say that a dataset contains a column called Height. Height column has short, medium and tall labels. Label encoding is used to transform the height column into 0, 1 and 2. i.e., the labels for short, medium and tall heights are 0, 1 and 2 respectively, to convert the categorical values into the numerical data. The process of converting labels into a numeric format so that machines can read them is known as label encoding. The sample dataset after preprocessing is shown in the Table 3. The data used in this paper was downloaded from Kaggle (https://www.kaggle.com/code/mdejazulhassan/vehicle-dataset-from-cardekho/data).

Table 1. Attribute description of the dataset

Name of the attribute	Description of the attribute
Car name	It provides the car's name
Year	The year when the car was bought is specified, ranging from 2003 to 2018
Selling price	It specifies the price charged so that the owner sells the car
Present price	It specifies the actual price of the vehicle outside the showroom
Kms driven	It gives the car's total distance travelled in kilometres
Fuel type	It specifies the car's fuel type
Seller type	It states whether the vendor is either a merchant or a person
Transmission	It specifies if the vehicle is automatic/ manual
Owner	It identifies former owners of a vehicle

Table 2. Sample dataset before preprocessing

Car name	Year	Selling price	Present price	Kms driven	Fuel type	Seller type	Transmission	Owner
ritz	2014	3.35	5.59	27000	Petrol	Dealer	Manual	0
sx4	2013	4.75	9.54	43000	Diesel	Dealer	Manual	0
ciaz	2017	7.25	9.85	6900	Petrol	Dealer	Manual	0
wagon r	2011	2.85	4.15	5200	Petrol	Dealer	Manual	0
swift	2014	4.6	6.87	42450	Diesel	Dealer	Manual	0
vitarabrezza	2018	9.25	9.83	2071	Diesel	Dealer	Manual	0
ciaz	2015	6.75	8.12	18796	Petrol	Dealer	Manual	0
s cross	2015	6.5	8.61	33429	Diesel	Dealer	Manual	0
ciaz	2016	8.75	8.89	20273	Diesel	Dealer	Manual	0
ciaz	2015	7.45	8.92	42367	Diesel	Dealer	Manual	0
alto 800	2017	2.85	3.6	2135	Petrol	Dealer	Manual	0
ciaz	2015	6.85	10.38	51000	Diesel	Dealer	Manual	0
ciaz	2015	7.5	9.94	15000	Petrol	Dealer	Automatic	0
ertiga	2015	6.1	7.71	26000	Petrol	Dealer	Manual	0

Table 3. Sample dataset after preprocessing

S. No	Car Name_cat	Year	Present Price	KmsDriven	Fuel Type_cat	Seller Type_cat	Transmission_cat	Owner
0	90	2014	5.59	27000	2	0	1	0
1	93	2013	9.54	43000	1	0	1	0
2	68	2017	9.85	6900	2	0	1	0
3	96	2011	4.15	5200	2	0	1	0
4	92	2014	6.87	42450	1	0	1	0
...	...	...	...	...	...	...	...	...
296	69	2016	11.60	33988	1	0	1	0
297	66	2015	5.90	60000	2	0	1	0
298	69	2009	11.00	87934	2	0	1	0
299	69	2017	12.50	9000	1	0	1	0
300	66	2016	5.90	5464	2	0	1	0

4 Methodology

The proposed workflow diagram is shown in the Fig. 1. There are six steps and these are described below.

Step 1: Data Collection – The dataset is collected from the Kaggle.
Step 2: Data Processing – By eliminating every row with a null value, the dataset is cleaned. Categorical values are present in the dataset for the attributes Car Name, Fuel Type, Seller Type, and Transmission. Using label encoding, these categorical variables are transformed into numerical values. Label coding is the process of converting labels into a numeric format that can be read by computers.
Step 3: Applying ML Algorithms for Analysis – An 80:20 ratio is used to divide the datasets into training and test data. In order to provide a flexible and consistent dataset, the data has been standardized. This was accomplished using Standard Scaler from the Scikit-Learn Library. It normalizes the features by deleting the mean and scaling the unit variance. After normalizing the range of features in the datasets, next is to forecast the selling price of used cars, a variety of techniques are used, such as CatBoost Regression, Random Forest Regression, and Support Vector Regression.

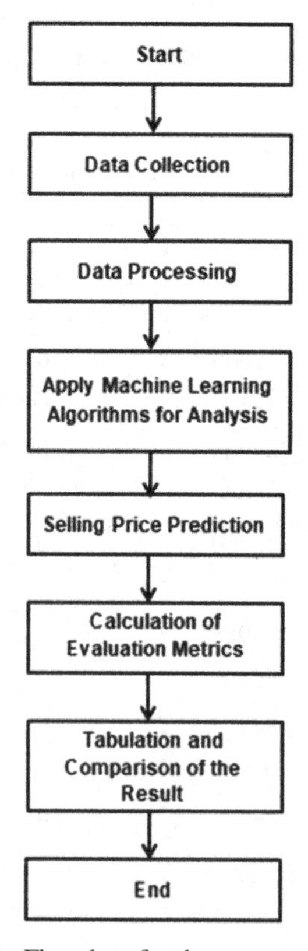

Fig. 1. Flowchart for the proposed work

Step 4: Selling Price Prediction – The accuracy of the used car selling price is estimated with the help of machine learning techniques.
Step 5: Calculation of Evaluation Metrics – R-SQUARE, MSE, RMSE, MAE, and the accuracy (1-MAE) of CatBoost Regression, Random Forest Regression, and Support Vector Regression are the metrics employed for the proposed work. The accuracy formula is shown in eq. (1).

$$Accuracy\ (A) = 1 - Mean\ Absolute\ Error\ (MAE) \tag{1}$$

Step 6: Tabulation and Comparison of the Result – Taking into account the accuracy of the algorithms as well as R-SQUARE, MSE, RMSE, and all other metrics that were acquired after using machine learning techniques. A number of graphs, including line graphs, density plots, and scatter plots, are plotted for each model for the purpose of comparison tabulating. All metric values for each model are further tabulated, and bar graphs are generated to compare the accuracy of each model as well as the values for R-SQUARE, MSE, RMSE, and MAE. The next step after tabulating all the numbers is to compare the metric values of all the used algorithms to determine which one best fits the scenario.

5 Result and Discussion

The proposed work uses the inbuilt Python libraries such as NUMPY, PANDAS (for linear algebra and scientific computing) and visualization libraries such as MATPLOTLIB, SEABORN. It uses the Label Encoder to convert all the other data types to numeric data type.

The closeness between predicted and actual data CatBoost Regression is shown in the Fig. 2. Density Plot to determine distribution of variables in the dataset is shown in Fig. 3. Figure 4 displays a scatter plot to show the correlation between the actual and expected values. The variance between actual and the predicted sales price (sample) is shown in the Table 4 and if it is accurate or not in the CatBoost Regression model. The evaluation metrics used in CatBoost Regression model are R2 score, MSE, RMSE, MAE and accuracy and their values are shown in the Fig. 5.

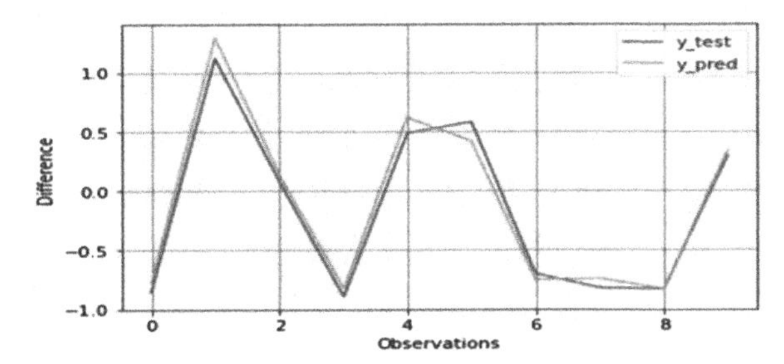

Fig. 2. Plot to determine the closeness between predicted and actual data using CatBoost Regression

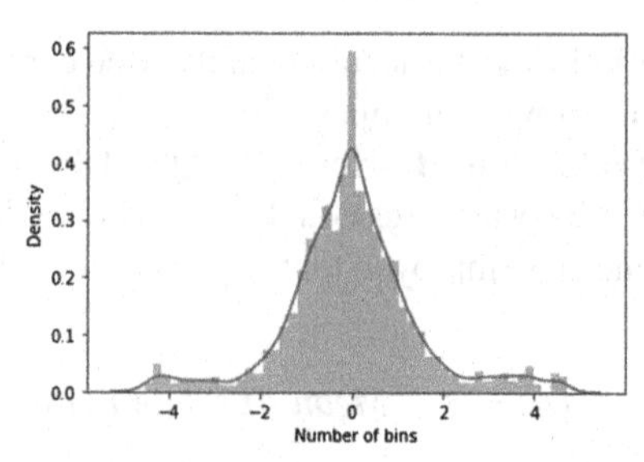

Fig. 3. Density Plot to determine distribution of variables in the dataset using CatBoost Regression

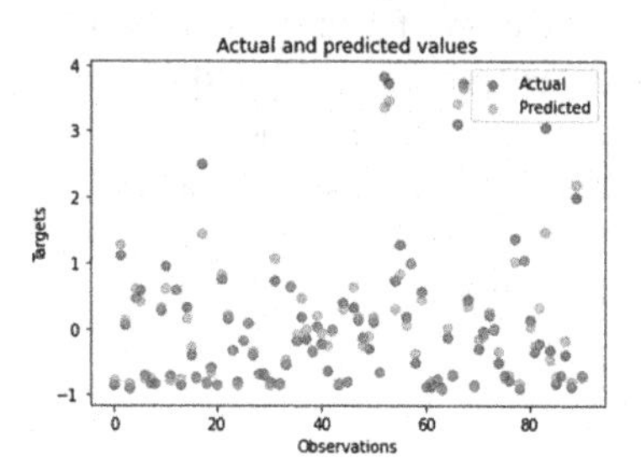

Fig. 4. Scatter plot to determine correlation between actual and predicted values using CatBoost Regression

Table 4. The variance between actual and the predicted sales price (sample) and if it is Accurate or not in the CatBoost Regression model

S. No	Actual selling price	Predicted selling price	Difference	ACC or not	Percentage error
0	−0.852481	−0.773279	−0.079203	Accurate	−9.290831
1	1.117623	1.291965	−0.174343	Accurate	15.599416
2	0.076051	0.135253	−0.059202	Accurate	77.844952
3	−0.892852	−0.821449	−0.071404	Accurate	−7.997279
4	0.479761	0.618706	−0.138945	Accurate	28.961353
5	0.580689	0.414865	0.165824	Accurate	28.556397
6	−0.701090	−0.752449	0.051359	Accurate	−7.325561
7	−0.822203	−0.743478	−0.078726	Accurate	−9.574965
8	−0.832296	−0.832362	0.000066	Accurate	−0.007872
9	0.287999	0.336501	−0.048502	Accurate	16.841112
10	0.944028	0.618321	0.325706	Accurate	34.501768

```
R2 SCORE is 0.9438514248527179
mean_sqrd_error is  0.06514319729907371
Root mean squared error of is 0.255231654187081
Mean Absolute error is 0.1371290433308082
Accuracy is 0.8628709566691918
```

Fig. 5. Evaluation metrics values for CatBoost Regression model

The closeness between predicted and actual data Random Forest Regression is shown in the Fig. 6. Density Plot to determine distribution of variables in the dataset is shown in Fig. 7. Scatter plot to determine correlation between actual and predicted values are shown in Fig. 8.

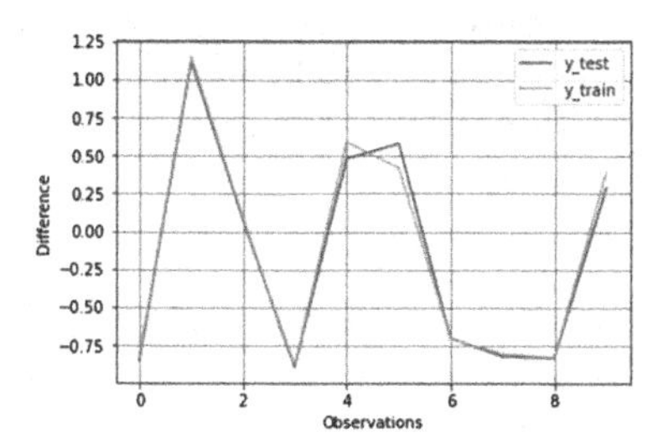

Fig. 6. Plot to determine the closeness between predicted and actual data using Random Forest Regression

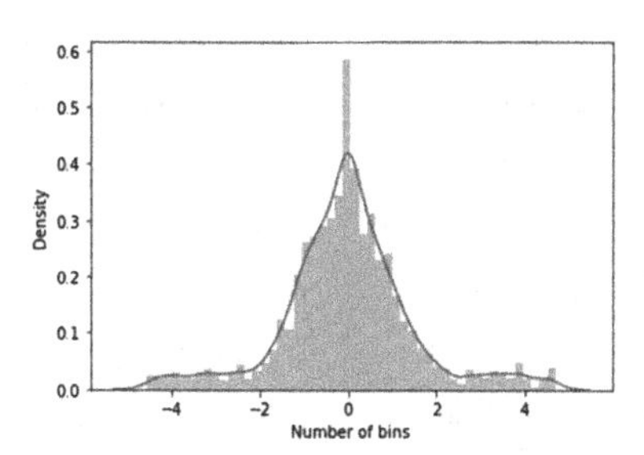

Fig. 7. Density Plot to determine distribution of variables in the dataset using Random Forest Regression

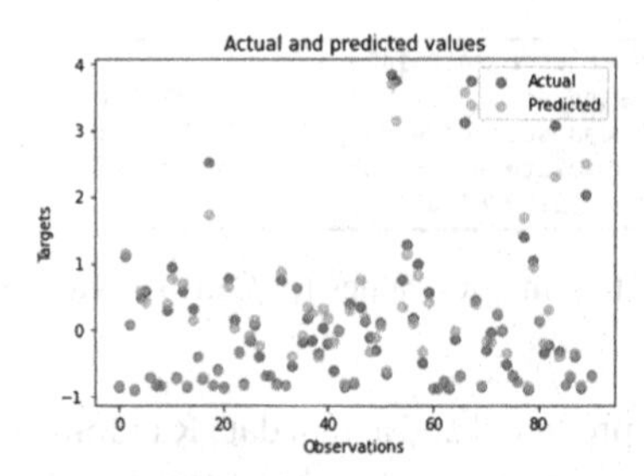

Fig. 8. Scatter plot to determine correlation between actual and predicted values using Random Forest Regression

The Table 5 shows the variance between actual and the predicted sales price (sample) and if it is accurate or not in the Random Forest Regression model. The evaluation metrics used in Random Forest Regression model are R2 score, MSE, RMSE, MAE and accuracy and their values are shown in the Fig. 9.

Table 5. The variance between actual and the predicted sales price (sample) and if it is accurate or not in the Random Forest Regression model

S. no	Actual selling price	Predicted selling price	Difference	ACC or not	Percentage error
0	−0.852481	−0.828221	−0.024261	Accurate	−2.845921
1	1.117623	1.151512	−0.033889	Accurate	3.032278
2	0.076051	0.092846	−0.016794	Accurate	22.082912
3	−0.892852	−0.880713	−0.012140	Accurate	−1.359637
4	0.479761	0.591234	−0.111472	Accurate	23.234974
5	0.580689	0.420214	0.160475	Accurate	27.635238
6	−0.701090	−0.705139	0.004049	Accurate	−0.577559
7	−0.822203	−0.805542	−0.016661	Accurate	−2.026398
8	−0.832296	−0.828136	−0.004160	Accurate	−0.499850
9	0.287999	0.395880	−0.107881	Accurate	37.458948
10	0.944028	0.769627	0.174401	Accurate	18.474107

```
R2 SCORE is 0.9621157252658848
mean_sqrd_error is  0.04395307943368595
Root mean squared error of is 0.20964989728994848
Mean Absolute error is 0.1283807660158989
Accuracy is 0.8716192339841011
```

Fig. 9. Evaluation metrics values for random forest regression model

The closeness between predicted and actual data Support Vector Regression (SVR) is shown in the Fig. 10. Density Plot to determine distribution of variables in the dataset using SVR is shown in Fig. 11. Figure 12 displays a scatter plot using SVR to show the correlation between the actual and expected values. The variance between actual and the predicted sales price (sample) is shown in the Table 6 and if it is accurate or not in the Support Vector Regression model. The evaluation metrics used in Support Vector Regression model are R2 score, MSE, RMSE, MAE and accuracy and their values are shown in the Fig. 13.

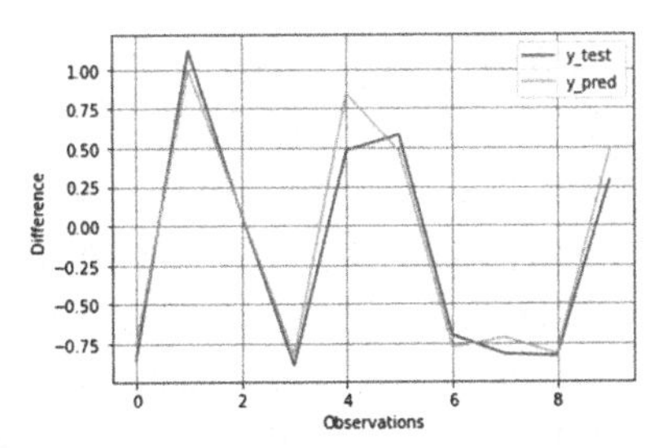

Fig. 10. Plot to determine the closeness between predicted and actual data using Support Vector Regression (SVR)

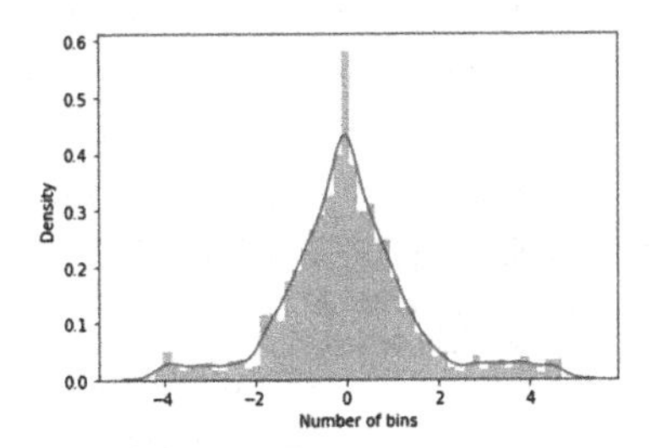

Fig. 11. Density Plot to determine distribution of variables in the dataset using SVR

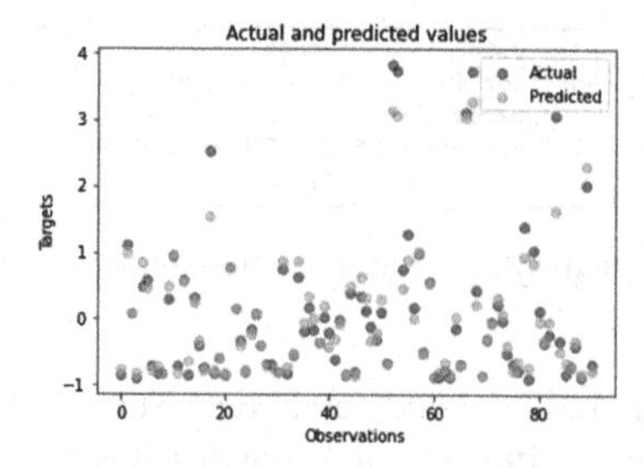

Fig. 12. Scatter plot to determine correlation between actual and expected values using SVR

Table 6. The variance between actual and the expected sales price (sample) and if it is Accurate or not in the SVR model

S. No	Actual selling price	Predicted selling price	Difference	ACC or not	Percentage error
0	−0.852481	−0.760063	−0.092418	Accurate	−10.841074
1	1.117623	0.995251	0.122372	Accurate	10.949334
2	0.076051	0.081998	−0.005947	Accurate	7.819079
3	−0.892852	−0.827224	−0.065628	Accurate	−7.350375
4	0.479761	0.843538	−0.363777	Accurate	75.824640
5	0.580689	0.474667	0.106022	Accurate	18.257982
6	−0.701090	−0.779501	0.078411	Accurate	−11.184135
7	−0.822203	−0.715066	−0.107138	Accurate	−13.030565
8	−0.832296	−0.818867	−0.013429	Accurate	−1.613529
9	0.287999	0.479512	−0.191513	Accurate	66.497896
10	0.944028	0.954947	−0.010920	Accurate	1.156733

```
R2 SCORE is 0.9453731791724593
mean_sqrd_error is  0.06337766822497694
Root mean squared error of is 0.25174921693021596
Mean Absolute error is 0.14590172524212977
Accuracy is 0.8540982747578703
```

Fig. 13. Evaluation metrics values for SVR model

Comparison of various measures like R2 score, MSE, RMSE, MAE and accuracy using Support Vector Regression, Random Forest Regression, and CatBoost Regression and their values are shown in the Table 7. When compared to Support Vector Regression and CatBoost Regression, Random Forest Regression has the highest accuracy of 87.16% and it is shown in the Fig. 14. Random Forest Regression gives the lowest MAE of 0.1284 when compared to Support Vector Regression and CatBoost Regression and it is shown in the Fig. 15.

Table 7. Comparison of various measures used in various algorithms

	Support vector regression	Random forest regression	CatBoost regression
R2	0.9454	0.9621	0.9439
MSE	0.0634	0.044	0.0651
RMSE	0.2517	0.2096	0.2552
MAE	0.1459	0.1284	0.1371
Accuracy (%)	85.4	87.16	86.28

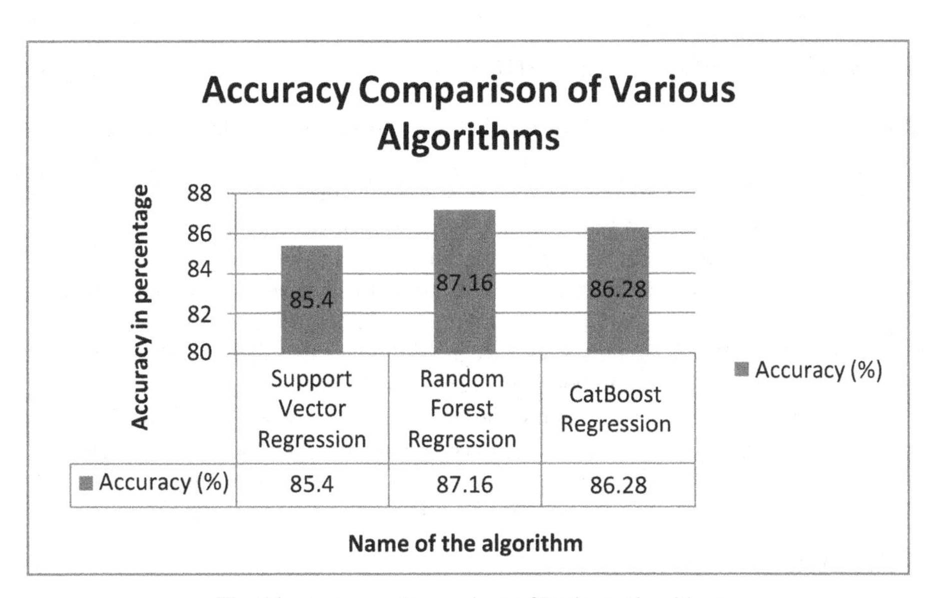

Fig. 14. Accuracy Comparison of Various Algorithms

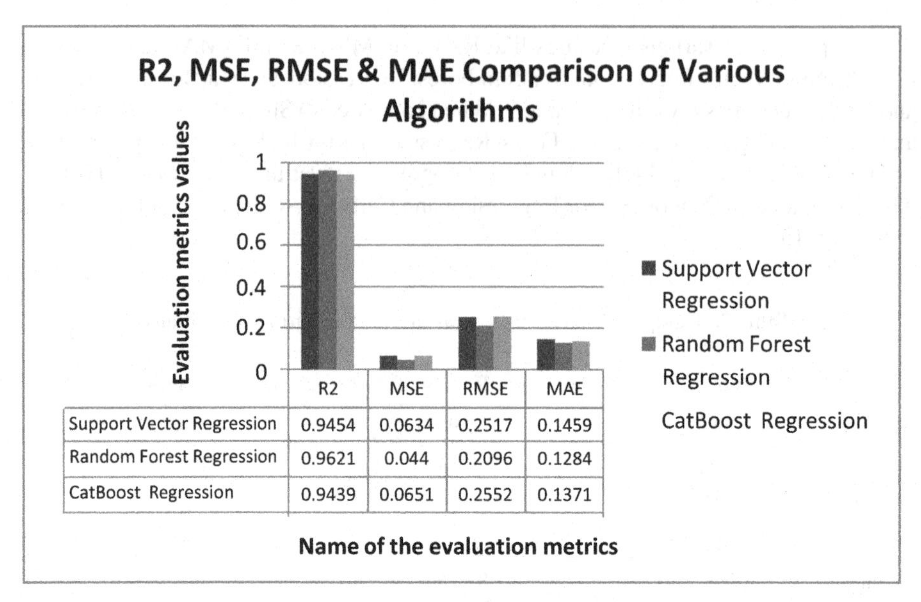

	R2	MSE	RMSE	MAE
Support Vector Regression	0.9454	0.0634	0.2517	0.1459
Random Forest Regression	0.9621	0.044	0.2096	0.1284
CatBoost Regression	0.9439	0.0651	0.2552	0.1371

Fig. 15. R2, MSE, RMSE & MAE Comparison of Various Algorithms

6 Conclusion and Future Work

This article uses three different machine learning methods to predict the cost of used cars. In the existing method [13], mean absolute error for decision tree regression was 0.6711, which was the least used among the various algorithms that they employed, including linear regression, Lasso regression, Ridge regression, Bayesian Ridge regression, and others. The proposed system concludes that the Random Forest regression performs better than Support Vector Regression and CatBoost Regression. When compared to Support Vector Regression and CatBoost Regression, Random Forest Regression has the best accuracy (87.16%). Support Vector Regression gives the highest MAE of 0.1459 when compared to and Random Forest Regression and CatBoost Regression.

In this study, the primary problem is the little number of records used. We intend to collect additional data and use more advanced techniques, such as fuzzy logic, genetic algorithms, and deep learning approaches to forecast car costs in the future.

References

1. Van Thai, D., Ngoc Son, L., Tien, P.V., Nhat Anh, N., Ngoc Anh, N.T.: Prediction car prices using quantify qualitative data and knowledge-based system. In: 2019 11th International Conference on Knowledge and Systems Engineering (KSE), Da Nang, Vietnam, pp. 1–5 (2019)
2. Sameerchand, P.: Predicting the price of used cars using machine learning techniques. Int. J. Inform. Comput. Technol. **4**, 753–764 (2014)
3. Fathalla, A., Salah, A., Li, K.: Deep end-to-end learning for price prediction of second- hand items. KnowlInf Syst. **62**, 4541–4568 (2020)

4. Xia, Z., Xue, S., Wu, L.: ForeXGBoost: passenger car sales prediction based on XGBoost. Distrib. Parallel Databases **38**, 713–738 (2020)
5. Viswapriya, S.E., Durbaka, S., Sandeep, S., Gandavarapu, S.K.: Vehicle price prediction using SVM techniques. Int. J. Innovative Technol. Exploring Eng. **9**, 398–401 (2020)
6. Aashay, P.H., Saurabh, S.: Inspecting & cost evaluation of pre-owned cars using deep learning. Int. Res. J. Eng. Technol. **7**, 7578–7582 (2020)
7. Chuancan, C., Lulu, H., Cong, X.: Comparative analysis of used car price evaluation models. In: 2017 AIP Conference Proceedings, vol. 1839, pp. 1–7 (2017). https://doi.org/10.1063/1.4982530
8. Sun, N., Bai, H., Geng, Y., Shi, H.: Price evaluation model in second-hand car system based on BP neural network theory. In: 2017 18th IEEE/ACIS International Conference on Software Engineering, Artificial Intelligence, Networking and Parallel/Distributed Computing (SNPD), pp. 431–436 (2017). https://doi.org/10.1109/SNPD.2017.8022758
9. Han, S., Qu, J., Song, J., Liu, Z.: Second-hand car price prediction based on a mixed-weighted regression model. In: 2022 7th International Conference on Big Data Analytics (ICBDA), pp. 90–95 (2022). https://doi.org/10.1109/ICBDA55095.2022.9760371
10. Yu, L., Yan, H., Liu, L., Yi, S., Zhao, Y.: Car trading cycle prediction based on random forest algorithm. In: 2022 7th International Conference on Intelligent Computing and Signal Processing (ICSP). Pp. 639–642 (2022). https://doi.org/10.1109/ICSP54964.2022.9778603
11. Chen, S., Liu, Z.: Application of data mining technology in second-hand car price forecasting. In: 2022 3rd International Conference on Electronic Communication and Artificial Intelligence (IWECAI), pp. 260–273 (2022). https://doi.org/10.1109/IWECAI55315.2022.00058
12. Satapathy, S.K., Vala, R., Virpariya, S.: An automated car price prediction system using effective machine learning techniques. In: 2022 International Conference on Computational Intelligence and Sustainable Engineering Solutions (CISES), pp. 402–408 (2022). https://doi.org/10.1109/CISES54857.2022.9844350
13. Sharma, A.D., Sharma, V., Mittal, S., Jain, G., Narang, S.: Predictive analysis of used car prices using machine learning. Int. Res. J. Modernization Eng. Technol. Sci. **3**, 674–684 (2021)

Machine Learning Framework for Identification of Abnormal EEG Signal

A. Prabhakara Rao(✉), J. Bhaskar, and G. Prasanna Kumar

Department of ECE, Vishnu Institute of Technology, Bhimavaram, AP, India
prabhakararao.a@vishnu.edu.in

Abstract. Epilepsy is a chronic, noncommunicable disease (NCD) causing disorder in the brain's neurological activity. It may be due to genetic disorder or brain injuries caused by some accidents. This may cause seizures, loss of awareness, unusual sensations and behavior. Globally 50 million people are suffering from epilepsy, so it is one of the most prominent neurological diseases globally according to World Health Organization (WHO) statistics in 2021. It is estimated that up to 70% of people are suffering with epilepsy and they can be saved by timely diagnosis and proper treatment of epilepsy. Electroencephalograms (EEGs) are universally used to detect this chronic non-communicable disease. Furthermore, assessing a specific type of abnormality by visual examination of an EEG signal is an intuitive process that can vary from radiologist to radiologist. It is a challenging task for the radiologists to visually examine the EEG signal by looking for a shift in frequency or amplitude in long-duration signals. It may give rise to inaccurate categorization. Identification of epileptic seizure from the recorded EEG signal is a primary task in the treatment of epilepsy. In this work, wavelets were used to obtain the appropriate features from EEG signals. These features were fed to different classifiers. This work proposes a machine learning (ML) framework to detect the abnormality in the EEG signal automatically to assist the radiologists in their diagnosis. The ML framework uses 7 classifiers (KNN, SVM, Random Forest, Logistic Regression, Decision Tree, AdaBoost, and Bagging). Among these classifiers, Bagging Classifier was shown better performance in terms of accuracy and ROC.

Keywords: Electroencephalogram (EEG) · Epilepsy · Seizure · Wavelets · Machine learning (ML) framework · ML classifiers/

1 Introduction

Digital image and signal processing applications are spread in many fields [1, 2], among these, the application of image processing and signal processing in the medical field is a trending technique [3] in the current scenario. According to World Health Organization (WHO) estimates from 2021, epilepsy is one of the most prevalent neurological illnesses worldwide. Approximately 50 million people will be impacted by it, according to WHO figures until 2021 [4]. It results in muscle stiffness, seizures, etc. Prolonged seizures

N. Gupta et al. (Eds.): IC4S 2022, LNICST 472, pp. 42–54, 2023.
https://doi.org/10.1007/978-3-031-28975-0_4

can harm the brain. The brain's ability to operate will probably be negatively affected by isolated, brief seizures, and some brain cells may even be lost. More than half of the population suffering from epilepsy could live seizure-free lives if the condition was adequately recognized and treated in its early stages. It is a severe neurological condition with distinctive traits that is prone to recurrent seizures. This illness affects all mammal species, including rats, dogs, and cats in addition to people. However, the term "epilepsy" is unremarkable and consistently distributed around the world; it offers no hints as to the kind or severity of the seizures [5]. The classification of seizures is shown in Fig. 1.

Based on the symptoms and signs, seizures are classified into two major groups – focal and generalized [6]. Focal seizures affect one side of the brain (hemisphere) and the patient may lose consciousness. Partial or focal seizures are classified into simple and complex. Simple focal seizures are characterized by staring spells, automatisms, and sensory phenomena. Complex focal seizures involve confusion and disorientation. Generalized seizures influence both hemispheres affecting both the sides of the brain simultaneously and are accompanied by tonic-clonic movements. These include absence, myoclonic, tonic, and tonic-clonic seizures [7]. Absence seizures are characterized by sudden loss of consciousness. Myoclonus refers to jerking movements. Tonic seizures are characterized by stiffening of muscles. Tonic-clonic seizures may lead to rhythmic contractions of muscles. Generalized seizures basically classified into two categories namely convulsive and non-convulsive.

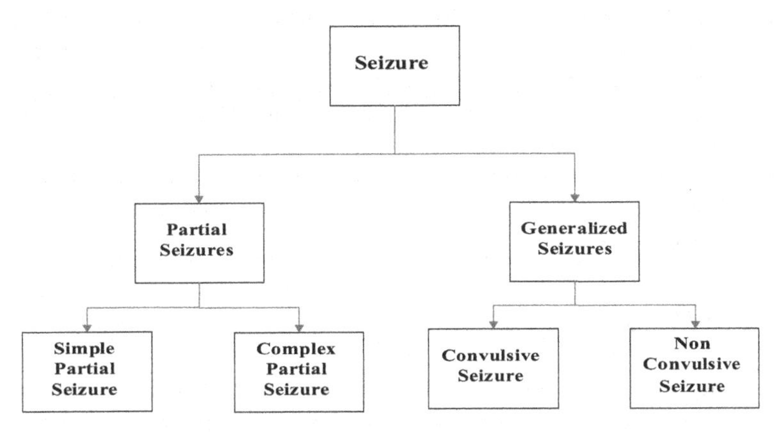

Fig. 1. Classification of seizures

Various non invasive approaches such as functional magnetic resonance imaging (fMRI), positron emission tomography (PET), nuclear magnetic resonance fluorescence, single-photon emission computed tomography, near-infrared spectroscopy, electroencephalogram (EEG) are commonly used to study the function of the brain. Among all these techniques, EEG is widely used because it records the electrical behaviors of the brain very accurately. Also, EEG is a simple, safe and less painful test. In diagnostic applications, priority is frequently given to spectral data. Electroencephalography (EEG), a method of electrophysiological monitoring, captures the brain's electrical activity. The electrodes are typically non-invasive and placed throughout the scalp, while occasionally invasive electrodes are used in specific applications. EEG records voltage changes in the brain's neurons due to ionic current. In therapeutic settings, an EEG is a long-term recording of the electrical activity of the brain made with many electrodes positioned on the scalp [8]. Electroencephalograms (EEGs) are universally used to detect epilepsy. Assessing a specific type of abnormality by visual examination of an EEG signal is an intuitive process that can vary from radiologist to radiologist. It is a challenging task for the radiologists to do visual examination of the recorded EEG signal and identify the shift in frequency or amplitude in the EEG signals of long-duration. It may give rise to inaccurate categorization. Determination of epileptic seizure is an essential task in the treatment of epilepsy. In this work, wavelets were used to obtain the appropriate features from EEG signals. Among the various signal processing techniques, wavelet transforms have the ability to efficiently identify the subtle changes in the EEG signal [9]. The wavelet features were fed to different classifiers. This paper proposes a machine learning (ML) framework consisting of 7 ML classifiers to detect the abnormality in the EEG signal automatically to assist the radiologists in their diagnosis.

The work presented in this paper is organized in the following manner. Section 1 gives brief introduction to Epileptic Seizures, EEGs, and the objective of the presented work. Section 2 deals with some of the related work. The methodology of the work is presented in Sect. 3. The results and discussions are included in Sect. 4, and the conclusions are described in Sect. 5.

2 Related Works

Several studies have been carried out to develop a system that can reliably identify abnormal EEGs in humans. This is because epilepsy [10], sleep disorders [11], and other conditions may be identified if these EEGs are correctly classified. These studies give equal attention to seizure detection and seizure prediction. The EEG signals which are non-linear in nature and are dynamic are difficult to analyze through linear techniques to produce consistent, accurate results. As a result, alternative machine learning (ML) or deep learning (DL) methodologies are applied in diverse investigations. Different features are extracted in the ML investigations. Wavelet transform, Hilbert-Huang transform, Eigen value decomposition, higher-order spectra, and cumulate features are some of the common examples. Depending on the study, one channel or multi-channel signals may be used. Different classifiers are employed in ML to categorize signals based on their signature or extracted features, such as k-nearest neighbor (KNN), support vector machine (SVM), random forest (RF), bagged trees, etc. Component analysis can also be used to categorize EEGs, in addition. Principal component analysis (PCA) was performed by Lopez et al. [12] with the KNN and RF classifiers, yielding accuracy rates of 58.2% and 68.3%, respectively. After pre-processing the data to remove various artifacts and noise, signal processing tasks begin with normalizing of the signals. The attributes are extracted after pre-processing. The collected attributes are then supplied to the classifiers, and the effectiveness of the classification is evaluated. If a model performs as expected, it is put to further test using a fresh, unrelated set of data. Studies presently use a variety of deep learning-based techniques that don't necessitate feature extraction and selection. Using the same database and a one dimensional convolutional neural network (1D-CNN) with single-channel signals lasting 60 s, Yildirim et al. [13] identified the aberrant EEG signals and discovered an error rate of 20.6%. Diego et al. [14] proposed a system that combined 2D CNN with ML on four-channel signals and achieved an error rate of 21.2% in the detection of abnormal EEG signals. Acharya et al.[10] suggested a CNN-based method for automatically distinguishing between seizure and non-seizure EEG patterns (13 layers). 300 signals from 5 patients were employed in the investigation, and the classification accuracy was 88.67%. A 13-layer CNN model was used by Oh et al. [15] to provide a strategy for the detection of Parkinson's disease (PD). They were able to attain an accuracy of 88.25% in their investigation by using EEG data from 20 healthy people and 20 patients with Parkinson's disease. Existing works on automatic identification of abnormal EEG signal using deep learning (DL) methods showed better accuracy but the computational load and memory requirements are high. Objective of the proposed work in this paper is to identify a suitable classifier and set of wavelet features that will identify abnormal EEG signal efficiently with minimum time and minimum computational load.

3 Methodology

This section describes the various steps involved in the implementation of the proposed machine learning framework for identification of abnormal EEG signal. The entire procedure was illustrated in a flowchart as shown in Fig. 2.

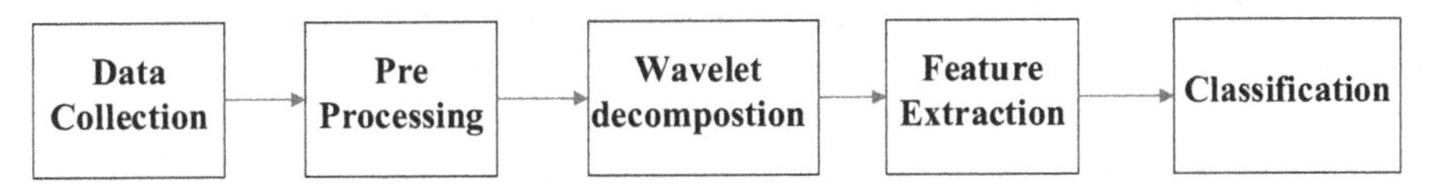

Fig. 2. Methodology of the proposed work

3.1 Data Collection

Epileptic seizure EEG data was collected from the Bonn dataset, an open-source data repository. The process of how these EEG recordings were obtained was explained in [16]. This data set contains five folders, each of which has 100 files that each represents a different subject or individual. An EEG recording of brain activity for duration of 23.6

s was stored in each file. The relevant time series 4097 data points are sampled in this analysis. As a result, we have collected 4097 data points across 23.5 s for a total of 500 candidates. Every 4097 data points were randomly divided into 23 pieces for each of the 178 sets of data that made up each chunk. Each data point displays the value of the EEG recording at a particular moment. Thus, we get $23 \times 500 = 11\ 500$ bits of information, each of which contains 178 datasets for a time of one second, with the last column 179 designating the response label y: 1, 2, 3, 4, and 5. The response variable (y) indicates the conditions under which the patient's EEG signal was recorded. Condition-5 (eyes open): The patient's eyes were open while the brain's EEG data was being captured, Condition-4 (eyes closed): The patient's eyes were closed while the EEG signal was being recorded, Condition-3 (tumor located, EEG in the normal area): tumor location in the brain was determined, and the healthy brain area activity was recorded through the EEG signal. Condition-2 (tumor area): brain activity in the tumor region. Condition-1: Seizure activity recording. Thus, the conditions-(1, 2, 3) and conditions-(4, 5) indicate unhealthy and healthy people, respectively. One healthy record and one unhealthy record which represent seizure activity were considered in this work from the given data.

3.2 Data Preprocessing

Initially the sampling frequency of the dataset is 178.3 Hz. It was resampled to 128 Hz frequency using bandpass filter and notch filter. All frequencies falling inside the passband would be sent to the output without being amplified or attenuated, whereas all frequencies falling outside the passband would be totally attenuated in an ideal bandpass filter. No bandpass filter is perfect in real life. As a result, we have filter roll-off. To eliminate this filter roll-off, notch filter was used as it blocks a specific band of frequencies and allows all frequencies outside the band.

3.3 Multilevel Wavelet Decomposition

Fourier Transforms, Fast Fourier Transform, Long Time Fourier Transform, Short Time Fourier Transform, Wavelet Transform, etc. can be used in the analysis of the EEG signal in frequency domain. But Wavelet transforms are highly efficient and robust while dealing with discrete signals. An orthogonal wavelet decomposition technique can be used to break down a signal into its component parts. A basic hierarchical framework is provided by a multi-resolution representation to examine the signal at various resolution levels. This is comparable to the idea of breaking down a signal into Walsh, Haar, or Fourier transform components. A signal is uniquely and entirely represented by its orthogonality. According to the Mallat theory, a signal's multi-resolution representation can be used to analyze its information content at various levels of detail [17]. A signal can be approximated by this operator at a specific resolution. Figure 3 depicts the wavelet's decomposition process used in the proposed work.

In the process of a wavelet transformation of a signal (S) is first decomposed into approximate coefficients and detailed coefficients. The signal's approximate [A1] (low frequency components) coefficient is the output of a low pass filter, and the signal's detailed [D1] (high frequency components) coefficient is the output of a high pass filter. This Approximate coefficient [A1] is again passed through a low pass to get approximate coefficient [A2] and A1 is passed through a high pass filter to obtain the detailed coefficient [D2]. Further A2 is decomposed into approximate coefficient [A3] and detailed coefficient [D3]. The number of decomposition levels depends on the length of signal and our requirements.

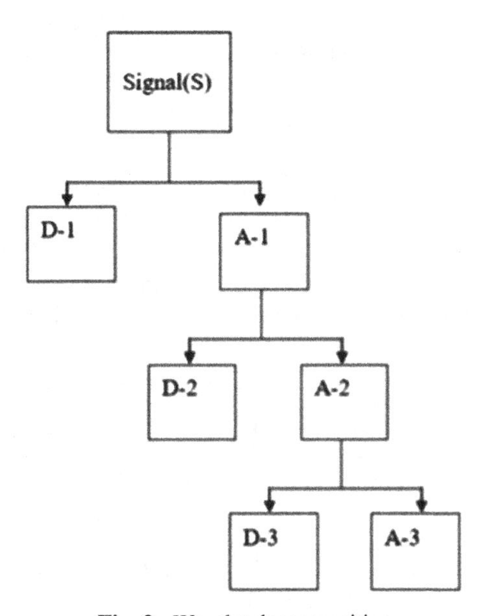

Fig. 3. Wavelet decomposition

The Original Signal S can be reconstructed with the help of A3, D3, D2 and D1.

With the decomposed wavelet coefficients the original signal can be reconstructed. The number of samples in next decomposition level is half as compared to previous stage. A1 and D1 will have N/2 samples if the original signal S had N samples, while A2 and D2 will have N/4 samples. The investigation of local signal behavior, such as spikes or discontinuities, is hence well suited for the wavelet transform. Because the frequencies change quickly and for a brief period at the site of discontinuity, we can investigate or analyze these abrupt shifts by selecting an appropriate time scale.

Db4 wavelet was applied on the two selected records to get the 4-level wavelet decomposition. As a result, 5 wavelet coefficients (a, d1, d2, d3, d4) were obtained for each record i.e., healthy and unhealthy. The dimensions of each coefficient record after applying the 4-level wavelet decomposition is 4098×100.

3.4 Feature Extraction

In this work, six features were calculated from each of the wavelet coefficient record obtained after wavelet decomposition. They are mean, variance, skewness, kurtosis, max_svd, entropy_svd. Definition and mathematical equation of the six features were illustrated below.

Mean

It is the ratio of the sum of all the compliances in the data to the total number of compliances. Therefore, the mean is a number surrounding which the entire data is spreading. It can be calculated as shown in Eq. 1.

$$\overline{x} = \frac{\sum_{i=1}^{N} x_i}{N} \tag{1}$$

where N represents the total number of observations and $\sum xi =$ sum of the observation

Variance

Variance measures how much variation there is within a group of data points. A low variance means that the data points are close together, while a high variance implies the data points are spread apart. It can be calculated using the formula shown in Eq. 2.

$$\sigma^2 = \frac{1}{N}\sum_{i=1}^{N}(x_i - \overline{x})^2 \tag{2}$$

Skewness

Asymmetry in a probability distribution function is measured as Skewness. It can be calculated using the formula shown in Eq. 3.

$$skewness = \frac{3(mean - median)}{standard\ deviation} \tag{3}$$

Kurtosis

Kurtosis describes the tagging of data whether it is lightly tagged, heavily tagged when compared to a normal distribution.

Max_svd

It is the maximum of singular value decompositions.

Entropy_SVD

It is a measure of the dimensionality of the data.

The SVD entropy of a signal X is defined as

$$H = \sum_{I=1}^{M} -(P_i * lnP_i) \tag{4}$$

where

- $P_i =$ normalized value of ith singular value of X,
- M = Total number of singular values in the embedded matrix X,

3.5 Classification of Abnormal EEG

In the proposed ML frame work 7 classifiers were used. They are namely Support Vector Machine (SVM), Decision Tree, K-Nearest Neighbor (KNN), Logistic Regression, Random Forest, AdaBoost, Bagging classifiers.

3.6 Performance Parameters

ML framework was built by using different classification models based on train data and predicts the results and compares them with test data using some parameters. They are Accuracy, Precision, Recallf1, Score, Confusion Matrix and ROC Curve.

Confusion Matrix
It serves as a performance indicator for classification problems using machine learning. It is a table containing four separate sets with actual and anticipated values. Recall, Precision, Specificity, Accuracy, and most critically area under the curve-receiver operating characteristic curve (AUC-ROC) are all very well measured by it. Accuracy and AUC are considered as performance measures in this work.

Accuracy
It describes the percentage of accurate predictions from the test records. It can be calculated from Eq. 5.

$$\text{Accuracy} = \frac{\text{TP} + \text{TN}}{\text{TP} + \text{TN} + \text{FP} + \text{FN}} \tag{5}$$

ROC Curve
Receiver operating characteristic (ROC) curve is a graphical representation of the performance of different classification models at all thresholds. It is plot of true positive rate versus false positive rate.

4 Results and Discussions

The classification report in terms of confusion matrix and RoC curve of the seven classifiers (SVM, Decision Tree, KNN, Logistic Regression, Random Forest, AdaBoost, and Bagging) were obtained and their performance was compared using accuracy and AUC-ROC. Confusion matrix and ROC Curve of the seven classifiers were shown in Figs. 4, 5, 6, 7.

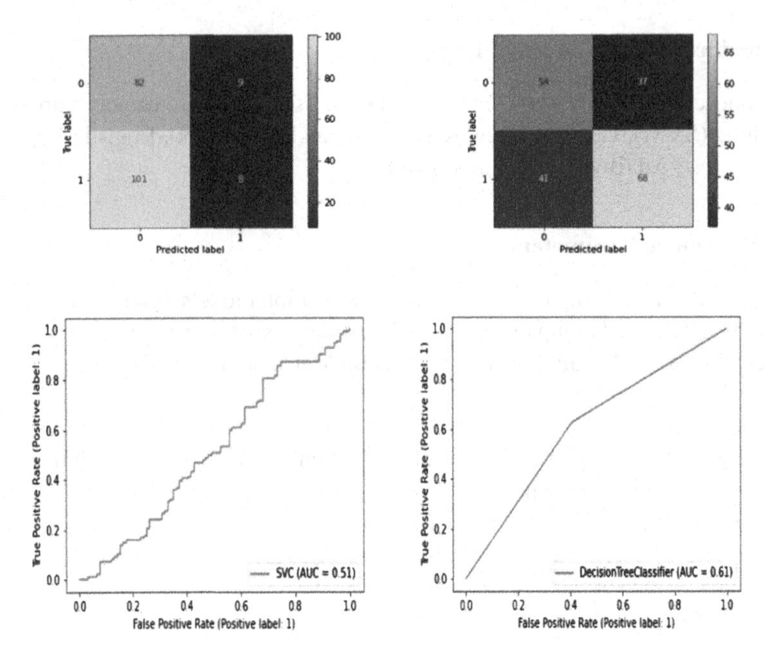

Fig. 4. Confusion matrix and ROC Curve of the SVM, Decision Tree Classifiers

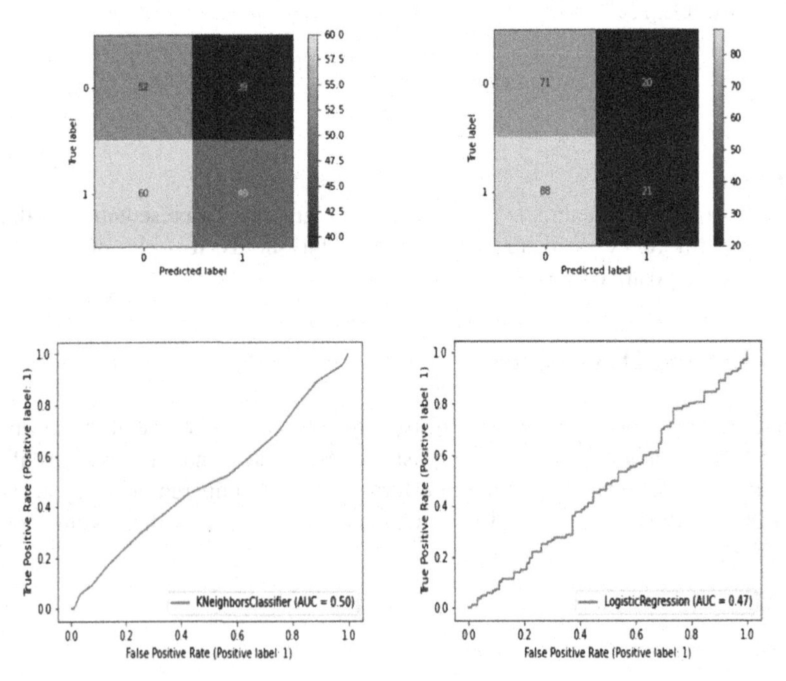

Fig. 5. Confusion matrix and ROC Curve of KNN and Logistic Regression Classifiers

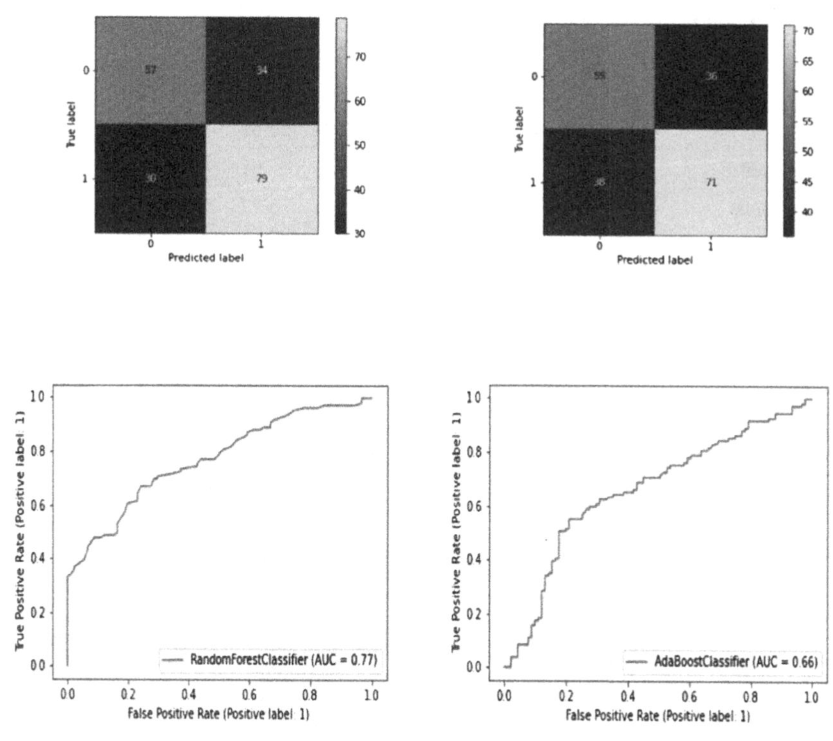

Fig. 6. Confusion matrix and ROC Curve of Random Forest and AdaBoost Classifiers

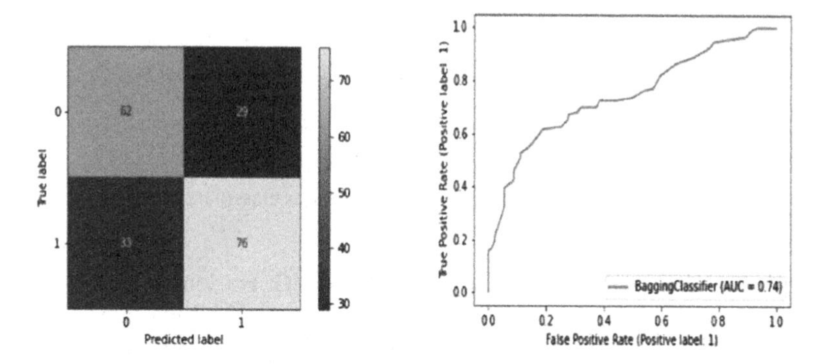

Fig. 7. Confusion matrix and ROC Curve of Bagging Classifier

Performance comparison of all the seven classifiers in terms of accuracy and % AUC-ROC were summarized in Table 1 and Fig. 8.

From Table 1 and Fig. 8, it was observed that among all the seven classifiers used in the ML framework, bagging classifier has shown better performance with a classification accuracy of 70% and AUC-ROC curve of 74%. This classifier can be used effectively in

Table 1. Performance of various classifiers used in the ML framework

Classifier	Accuracy (%)	AUC-ROC (%)
SVM	48	51
Decision Tree	65	61
Random Forest	68	77
AdaBoost	58	66
KNN	56	50
Logistic Regression	48	47
Bagging	70	74

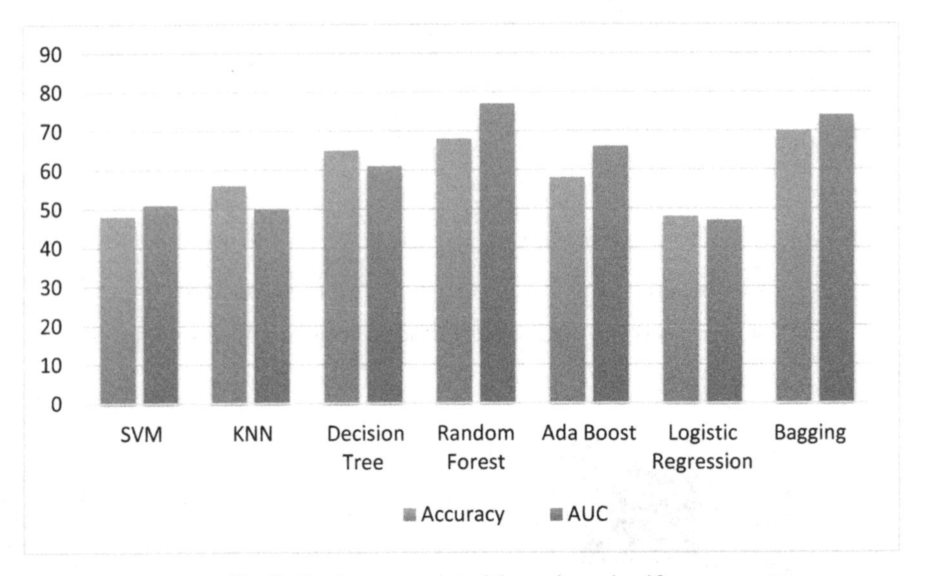

Fig. 8. Performance plot of the various classifiers

the detection of abnormal (seizure activity in EEG) EEG. For long-duration transmissions, it will be difficult for the clinicians to visually examine the EEG signal to identify the frequency or amplitude changes. Therefore the ML framework presented in this paper can be used in the automatic diagnosis of EEG signal. It improves the diagnostic accuracies as compared to the manual examination reducing the load on the clinicians.

This model's primary goal is to assist the radiologist in making an accurate diagnosis of EEG abnormalities. So, to categorize normal and abnormal EEG, which reflects seizure activity, an ML Framework with a variety of classification methods was used in the presented work. The EEG input was first fed into the ML framework model, which then underwent 4-level wavelet decomposition. The mean, variance, skewness, kurtosis, max_svd, and entropy characteristics are extracted from all the coefficients obtained after wavelet decomposition and then these features are given to several classification algorithms in the ML framework model, such as SVM, Decision Tree, KNN, Logistic Regression, Random Forest, AdaBoost, and Bagging. The bagging classifier in the proposed system performed well for the given EEG data with 70% accuracy and 76% AUC among all the classification algorithms. This framework model can be employed in a practical setup to analyze the EEG Signals in real time because it requires less computational resources and time as compared to the CNN models. Although the performance of the proposed model is less compared to CNN models it is good in terms of less computational load, memory requirement and time. In addition, performance of the ML framework presented in this work can be improved by analyzing on the suitable wavelet features among the five wavelet coefficients obtained from the four level wavelet decomposition. This ML framework can also be used to treat other anomalies, such as sleep disorders and other neurological disorders.

5 Conclusions

Information processing in the brain signal was recorded through EEG. Dynamic changes in the brain activity can be recorded through EEG which produces electrical signals varying in time, frequency and space. Various non linear and time-frequency analysis methods were used to analyze the EEG. Among these time-frequency analysis techniques, wavelet transforms were proven to be better as they efficiently capture the dynamic and subtle changes in the EEG signal. Therefore in the proposed work five wavelet features were extracted by using db4 wavelets on the EEG signal and these features were given as inputs to the seven classifiers to detect abnormal EEG. Performance of these classifiers was analyzed and it was observed that bagging classifier was showing better performance for the given EEG data with an accuracy of 70% and area under ROC curve of 74%. The proposed ML framework was used to automatically detect the abnormal EEG, which will be used to assist the clinicians. This approach can lighten the workload of the radiologist who is responsible for manually and visually identifying seizures on long-duration EEG signals.

References

1. Budumuru, P.R., Kumar, G.P., Raju, B.E.: Hiding an image in an audio file using LSB Audio technique. In: International Conference on Computer Communication and Informatics (ICCCI). IEEE (2021)
2. Sahu, S., Rao, A.P., Mishra, S.T.: Fingerprints based gender classification using adaptive neuro fuzzy inference system. In: International Conference on Communications and Signal Processing (ICCSP). IEEE (2015)
3. Rao, A.P., Bokde, N., Sinha, S.: Photoacoustic imaging for management of breast cancer: a literature review and future perspectives. Appl. Sci. **10**(3), 767 (2020)
4. World Health Organization: Epilepsy. Epilepsy Key facts 2022 [cited 2022 7 Oct 2022]. Available from: https://www.who.int/news-room/fact-sheets/detail/epilepsy
5. Reynolds, E.H., Rodin, E.: The clinical concept of epilepsy. Epilepsia **50**, 2–7 (2009)
6. Sharma, M., Patel, S., Acharya, U.R.: Automated detection of abnormal EEG signals using localized wavelet filter banks. Pattern Recogn. Lett. **133**, 188–194 (2020)

7. Mayoclinic: Eplilepsy overview (2021) [cited 7 Oct 2022]. Available from: https://www.mayoclinic.org/diseases-conditions/epilepsy/symptoms-causes/syc-20350093
8. HealthMedia: EEG (Electroencephalogram) overview. EEG Overview 2022, 9 Nov 2021 [cited 7 Oct 2022]. Available from: https://www.healthline.com/health/eeg
9. Faust, O., et al.: Wavelet-based EEG processing for computer-aided seizure detection and epilepsy diagnosis. Seizure **26**, 56–64 (2015)
10. Acharya, U.R., et al.: Deep convolutional neural network for the automated detection and diagnosis of seizure using EEG signals. Comput. Biol. Med. **100**, 270–278 (2018)
11. Michielli, N., Acharya, U.R., Molinari, F.: Cascaded LSTM recurrent neural network for automated sleep stage classification using single-channel EEG signals. Comput. Biol. Med. **106**, 71–81 (2019)
12. Lopez, S., et al.: Automated identification of abnormal adult EEGs. In: IEEE Signal Processing in Medicine and Biology Symposium (SPMB). IEEE (2015)
13. Yıldırım, Ö., Baloglu, U.B., Acharya, U.R.: A deep convolutional neural network model for automated identification of abnormal EEG signals. Neural Comput. Appl. **32**(20), 15857–15868 (2018). https://doi.org/10.1007/s00521-018-3889-z
14. Diego, S.L.d.: Automated identification of abnormal EEGs. In: Electrical Engineering, p. 63. Temple University (2017)
15. Oh, S.L., et al.: A deep learning approach for Parkinson's disease diagnosis from EEG signals. Neural Comput. Appl. **32**(15), 10927–10933 (2018). https://doi.org/10.1007/s00521-018-3689-5
16. Andrzejak, R.G., et al.: Indications of nonlinear deterministic and finite-dimensional structures in time series of brain electrical activity: dependence on recording region and brain state. Phys. Rev. E **64**(6), 061907 (2001)
17. Mallat, S.G.: A theory for multiresolution signal decomposition: the wavelet representation. IEEE Trans. Pattern Anal. Mach. Intell. **11**(7), 674–693 (1989)

Alzheimer's Disease Detection Using Ensemble of Classifiers

B. V. V. Satyanarayana, G. Prasanna Kumar(✉), A. K. C. Varma, M. Dileep, Y. Srinivas, and Prudhvi Raj Budumuru

Department of ECE, Vishnu Institute of Technology, Bhimavaram, Andhra Pradesh, India
godiprasanna@gmail.com

Abstract. Alzheimer's disease is a major intellectual deficit that makes it impossible for a person to carry out daily tasks. Finding the people with Alzheimer's and mild cognitive impairment is a difficult task. In order to arrange healthy, mildly cognitively impaired patients at the model stage itself using multimodal features. This paper will consider the presentation of cutting-edge Dynamic Ensemble Selection of Classifier computations. The review's data came from the Alzheimer's Disease Neuroimaging Initiative Dataset. For the purpose of expectation, the patients' clinical imaging, cerebrospinal fluid, cognitive test, and socioeconomic data are taken into consideration at the routine appointments. The demonstration of the most recent dynamic En-semble of Classifier Selection calculations is reviewed with the aid of these highlights in terms of Accuracy, Specificity and Sensitivity. Calculations for the Classifier Selection use the pool of machine learning classifiers that are used the most frequently as a contribution. Additionally, the display of the machine learning classifiers without using the computations for the Selection of Classifiers is also examined. Classifier selection calculations performed on the majority of the classifier pool to identify individuals with moderate cognitive impairment, Alzheimer's disease, and hearing loss have expanded presentation metrics including balanced classification accuracy, sensitivity, and specificity.

Keywords: Alzheimer's disease · Chronic illness · Clinical imaging · Machine learning · Dynamic ensemble classifier

1 Introduction

Digital Images processing is widely used in the fields of data security, image reconstruction, medical applications, etc.... [1–3]. Alzheimer's disease is a neurological condition that cannot be cured. The disarray of brain cells caused by AD is eradicating current affairs. Patients with AD will develop terrible memory problems as the disease progresses and lose their ability to carry out daily tasks. There is currently no clinical treatment that completely reverses AD. But early-stage pharmaceuticals may also redesign cues or lower the cost of memory loss. The four stages of the human intellectual framework

N. Gupta et al. (Eds.): IC4S 2022, LNICST 472, pp. 55–65, 2023.
https://doi.org/10.1007/978-3-031-28975-0_5

are included in the review. The specific levels linked to the early detection of AD are AD, CN, MCIc.

The retrieved functions from medical images are utilised to train the classifiers in custom characteristics-based methods. For specialists, the extraction of large hand-made capabilities is a laborious task. The development of deep models will allow for the instant extraction of capabilities from photos without the time-consuming efforts of a specialist. As a result, the deep learning trends are the focus of the research on disorder diagnosis. For the segmentation, detection, and classification of several brain illnesses in MRI images, deep learning models claimed outstanding results. The detection of AD from MRI scans has been proposed by several devices studying methods. The fundamental understanding of contemporary configurations using convolutional neural networks (CNN).

Figure 1 shows the diseased brain effected by Alzheimer's disease. We can observe that the Hippocampus is completely shrinked with enlarged ventricles and Tau neurofibrillary tangles. The above MRI images are taken from normal person and diseased person. The above two images (A, B) are explaining the healthy brain. The remaining two (C, D) are taken show the Alzheimer's disease. CNN played a significant capacity to facilitate the discovery of various sicknesses.

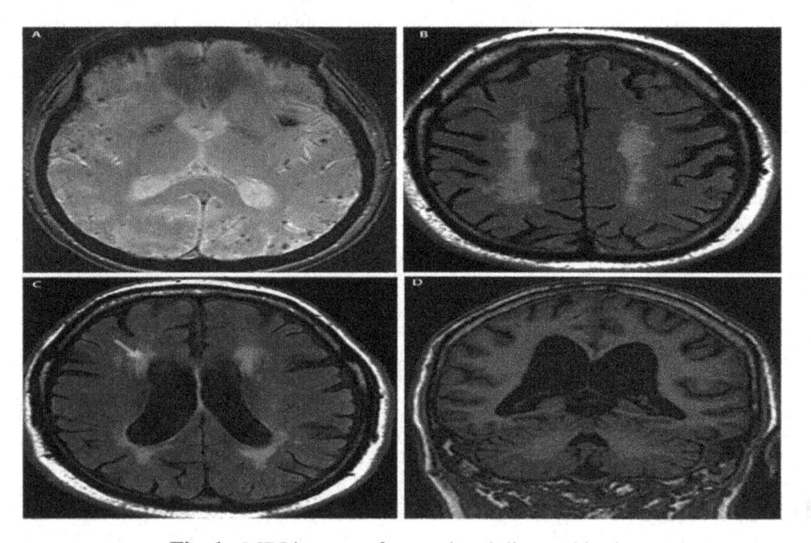

Fig. 1. MRI images of normal and diseased brain

It is accounted for that there are 50 million individuals are experiencing Alzheimer's illness. Furthermore, this choose is relied upon to skyrocket to 132 million by 2050. Individuals developed from at least 60 are encountering the Alzheimer's infection. So, in this concern we will recognize the Alzheimer's disease by using essential classifiers. This endeavour will help the experts in the field of medical industry to recognize stage of the Alzheimer's sickness.

- To Develop a Machine Learning system and provide medical sustainability for people who are suffering from Alzheimer's disease.
- Involving the fastest developing technology "Machine Learning" into public health and safety.
- To support up the precision levels from the current models.

Alzheimer's disease is a neurological condition that progresses, therefore the patient's ultimate prognosis is coma or death. Active therapies, however, can aid in slowing the process down.

The death of brain cells in Alzheimer's disease, a neurodegenerative illness, causes intellectual and cognitive impairment. It is the most common normal state of dementia, accounting for between 60 and 80% of all cases. There is currently no one research theory for AD; rather, specialists rely on a comprehensive medical history, cognitive assessments, computed tomography (CT) or magnetic resonance imaging (X-beam) scans of the brain, among other approaches, to confirm a suspicion. In general, there is no medication or technique that can stop the progression of AD.

However, for those who are experiencing mild or moderate stages of this condition, there are a few therapeutic drugs that can be used to quickly reduce symptoms and indicators and aid to enhance quality of life. Therefore, early AD investigation is quite important for routinely improved disease control. The investigations have shown a few links between neuropsychological assessments and brain degenerations. This can be used to verify AD early on in the separation process. In order to identify the most outrageously important capabilities in the investigation of AD, this paper uses brand name choice approaches [4].

This study demonstrates the necessity of combining MRI capabilities with neuropsychological results from the short mental state examination (MMSE) to improve the decisional space for an early assessment of AD. Exams like the MMSE nearby with different capabilities are discovered to redesign the request for orders without a doubt. The most crucial elements for the diagnosis of AD are extracted in this research using feature selection approaches. In order to improve the decisional space for an early diagnosis of AD, this article demonstrates the necessity of combining neuropsychological scores, such as those from the mini-mental state examination (MMSE), with MRI features. Through the experiments, it was discovered that adding the MMSE in addition to other variables considerably improved the classification of AD.

The most widely recognized kind of dementia, Alzheimer's disease, can trigger a neurological problem in the brain that results in mild cognitive decline by damaging synapses and the ability to carry out daily tasks. We can use Artificial Intelligence (AI) technology to identify and predict this illness using MRI (Attractive Resonance Imaging) filtered brain images, and we can organise AD patients according to whether they will or won't get this fatal illness in the future. The main reason for doing this is to create the greatest forecasting and identification tools for radiologists, specialists, and guardians to aid the patient who is suffering from this infection and save time and money [6].

The investigation of disease can now benefit greatly from deep learning (DL) methods due to their outstanding performance on large datasets. In this study, we implemented Convolutional Neural Network (CNN) for the earlier examination and planning of AD using the ADNI 3 class of MRI pictures, with unquestionably the quantity of 1512 delicate, 2633 common, and 2480 AD, to implement Convolutional Neural Network (CNN) for the earlier examination and plan of disease. The model performed well when compared to several other relevant researches, achieving a crucial exactness of almost 100%. Additionally, we compared the results to our earlier work in which machine learning calculations were applied using the Desert Spring dataset.

The deep learning methodologies can be a better choice than conventional AI techniques when handling massive volumes of information, such as clinical data. In order to identify Alzheimer's diseases on MRI images, CNN-based models are built with three distinct classifiers (SVM, NN, and FDT), and the model's performance was assessed between entirely connected layers. The objectives of the study are focused on the following research questions. Several traditional machine-learning techniques have been used in the past to study the diagnosis of Alzheimer's disease.

They are concentrating on creating models to examine the anatomical or structural brain images obtained by MRIs and the functioning of the brain in order to find any defects or abnormalities. Additionally, it viewed segmentation problems as classification problems and mainly relied on manually created features and feature representations for voxel, area, or patch-based approaches. To train classification algorithms, many expertly split images were needed, which requires more time. Recent years have seen significant advancements in the study of AD diagnoses and categorization employing DL methods. The findings demonstrate that this approach outperformed with 91.4% accuracy based on MRI and PET ADNI. However, this number drops to 82.6% when MRI is the only input and PET data is unavailable.

For patients' consideration and the development of treatments, it is important to detect Alzheimer's disease early on. This reality served as inspiration for the neuroimaging community, which successfully applied AI techniques to the early finding problem. The alliance of moves has helped the neighbourhood to resolve various difficulties that have been brought up and to normalise the approaches to handle the issue. In order to overcome the multiclass order problem, we use data from a global test for robotized forecast of MCI from MRI information [8]. By using paired t-test highlight selection, least squares multiclass subspace projection, and one to one error rectification yield codes arrangement; we offer a novel multiclass order technique that addresses the exception location problem.

In this paper, to evaluate the effectiveness of the recommended strategy and parameter fitting, two methods were used. To estimate the actual error on the training set, use re-substitution using the 10-fold cross validation approach. After the settings were adjusted, the accuracy was assessed using the test set. Based on the substitution estimation, a study was done regarding how to control the family-wise error (FWE) rate in our CAD system. 100 samples from HCs, 60 samples from the training set, and 40 samples from the test set made up the dataset (without dummies). Over the course of 1000 cycles, the dataset was randomly split into two subsets of 50 people each. The null hypothesis was then used to evaluate the re-substitution estimation. One PLS component (dimension) was chosen because there shouldn't be any group disparities in the feature set, and the real risk was set at 0.50. The obtained re-substitution accuracy had a standard deviation of 0.037 and was 0.612.

2 Methods for Detection of Alzheimer's

By using sophisticated neural organizations, numerous works are done to organize different stages of AD. To improve the decisional space for an early assessment of AD, it is necessary to combine neuropsychological results from tests like the Mini Mental State Examination with MRI capabilities. Through tests, such the MMSE nearby with varied skills, it has been discovered to upgrade the order of categories [7]. For the purpose of improving the order exactness between CN and MCIc, a troupe model is used. The outfit model consists of the pre-built organisation models MobileNet and Xception. With the use of the Alzheimer's infection neuroimaging drive (ADNI) dataset, the display of prepared and dressed models is tested.

The accuracy obtained using the MobileNet and Xception models is 89.23% and 89.89%, respectively. An accuracy of 91.3% in grouping is provided by the suggested model. The results demonstrate that the proposed outfit model has excellent ability to separate the stages MCIc and CN [5]. Convolutional Neural Network (CNN) was employed for the previous analysis and organization of AD using MRI images from the ADNI 3 class, with an absolute number of 1512 light, 2633 average, and 2480 AD.

Machine learning calculations using the Desert Spring dataset showed that deep learning algorithms can be more effective than traditional AI methods when dealing with large amounts of data, such as clinical data [6]. Mental indicators like age, the number of visits, the MMSE, and education are utilised in AI computations to forecast the beginning of Alzheimer's disease [7]. Paired t-test highlight selection, partial least squares multiclass subspace projection, and one-versus-one error rectification yield codes layout are used in a novel multiclass order technique that solves the issue of exception location. The proposed solution outperforms every alternative recommendation in the multiclass order with a precision of 67% [8]. AI algorithms to handle the advances in neuro imaging for early Alzheimer's disease detection [9]. An innovative approach is developed to identify MCI, Non-Convertible, and disease detection using hidden seductive resonation imaging (SMRI) from the challenge in Alzheimer's disease.

The method proposed in [10] is Support Vector Machine which is temporarily structured (TS-SVM) model be used in order to force revelation score for the missing MR picture development to increase monotonically with AD progress. Joint part decision and portrayal structure [11] gives the best morphological parts for attracting classifiers. The AI computation is used to execute the element extraction and component choice interaction, and the desert spring longitudinal dataset is then used for grouping [12, 13]. Results of the tests revealed that the SLR enhances the portrayal execution of AD/MCI when compared to other conventional methods [14].

It is obvious from the outcome of the division and depiction measures that the suggested framework outperforms other systems. This analysis shows that the suggested job is largely distinct and receives an MMSE score. The proposed pipeline correctly identified the HC region as the primary problem for diagnosing AD [15]. We used an underwriting algorithm to look at a huge subset of markers. Simply put, non-rude and frugally collectable markers are all that are required for the shift to AD in MCI and Pre MCI subjects [13]. Additionally, individual components and a mix of characteristics are determined as to how to individual classifiers and group-based classifiers in order to audit the introduction of the proposed technique. The accuracy attained using the combination of classifiers using the left hippocampus's size as a component is 92%, with 100% capability and 86% affectability.

3 Proposed Method Using Ensemble Classifier

The proposed approach contains four stages, that is, pre-processing, division, feature extraction, and portrayal of the MR pictures. It is displayed in the below Fig. 2. The core of our suggested method gets the information picture (MRI) ready for additional analysis. The image is divided into several regions in the second stage. In the third stage, features are isolated. The image appears to have a spot with an everyday topic or an Alzheimer's sufferer in the fourth and last stage. The graphic below shows how the suggested strategy actually functions. An ensemble is a collection of goods that are seen together rather than separately by utilizing this strategy in machine learning we may create models that in contrast to individual classifiers can generate the future data well. Multiple classifier systems, also known as ensemble systems, have attracted increasing interest from the computational intelligence and machine learning communities over the past couple of decades.

Any calculation used in image processing, such as element extraction and division, is entirely based on the characteristics of the images. The typical characteristic of Magnetic Resonance images degrades immediately after acquiring or shortly thereafter. Specific collectibles may be combined throughout the MRI acquisition process. For instance, force inhomogeneity and clamor issue. Different preprocessing techniques are necessary to remove them and update the image for additional review and analysis. In essence, picture preprocessing can build the image's visual stability. It deals with a variety of methods that enhance or eliminate some details in the image to manage it skillfully for future improvements. The proposed suggested method incorporates improvement change and averaging, power inhomogeneity repair, spatial standardization, and cerebrum extraction of surface in its preprocessing arrangement.

As they provide essential information to recognize healthy controls and Alzheimer's disease patients, neuro imaging assists best in early recognition of Alzheimer's contamination. However, the magnitude of the neuro images' enormous amount of information is the key problem here. The processing time required by the classifier to arrange these photographs is huge. Additionally, all of the image information is not necessary in order to characterization because a significant amount of it is unnecessary. As a result, feature extraction is carried out to take into account more relevant and discriminative components, to even more successfully depict photographs.

The correctness of order utilizing various sorts of items is assessed using SVM, Neural Networks, and Fine Decision Trees, three different types of classifiers. Additionally, we used a collection of these classifiers to assess the increased exactness rate. The data used in SVM classification might be linear or non-linear. An SVM Classifier allows for the setting of several kernels. We can designate the kernel as "linear" for a linear dataset. On the other hand, there exist two kernels for a non-linear dataset: "rbf" and "polynomial."

One of the wise look methodologies used in assessments, data mining, and AI is decision tree learning, or determination of call trees. It uses a decision tree (as an intelligent model) to get from observations of a few factors (tended to within the branches)

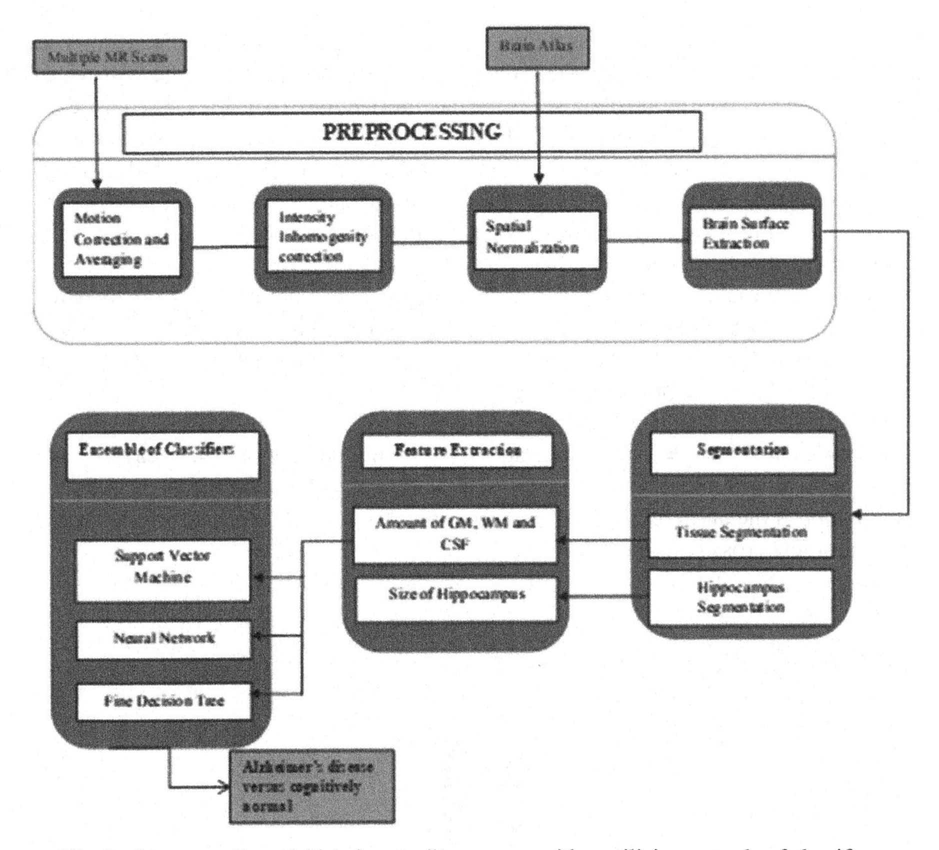

Fig. 2. Stream outline of Alzheimer's disease recognition utilizing cascade of classifiers.

to judgments about the object's objective value (tended to within the leaves). Collecting trees are tree models where the real variable will adopt a unique approach to attributes; in these tree structures, leaves address class checks and branches address part conjunctions that lead to those grouping names. As they fall away from the preferred trees, call trees are those places where the real part will take consistent attributes (often thoroughbred numbers). Given their comprehension and simplicity, call trees are among the most amazing AI evaluations.

4 Results and Discussion

The data set used in our project is Alzheimer's Disease Neuro-imaging Initiative Dataset. With help of the training models the inputs are given to the algorithm it classifies the dataset based on the classes and also produces the accuracy of detection as the iterations progresses. Here we test our algorithm with different data sets so as to see whether it is producing the correct output for every class. It detects the different classes among mildly cognitive, healthy, highly cognitive etc. and produces the output.

This section describes the test setup, which is followed by the results. The job is carried out through the use of MATLAB programming. Additionally, we have provided the machine with about 4,000 photos. During the information expansion procedure, each image in the underlying dataset is divided into 4 separate modifications. The data images are 3D MRI scans with several cuts. The central cut is separated from the 3D MRI image since processing with all cuts is monotonous and irrelevant.

The center cut is only taken into account for further processing. The preprocessing step has been standardization. Following preprocessing procedures, the order precision of MCIc against CN is observed using the accompanying pre-prepared organization models ResNet 50, VGG 16, Dense Net 121, Nas Net Mobile, Efficient NetB0, Xception, and MobileNet. The Xception and MobileNet models had the best characterisation accuracy among the pre-prepared models cited. Therefore, the proposed outfit model is created using the Xception and MobileNet models. Execution uses the following hyper-parameters: batch size = 7, learning rate = 0.001, momentum = 0.9, decay = 0.06, epochs = 100, iteration = 100, frequency = 50 iterations. The stochastic slope plummet computation is the analyzer used in this execution. The order misfortune work is based on the absolute cross-entropy task.

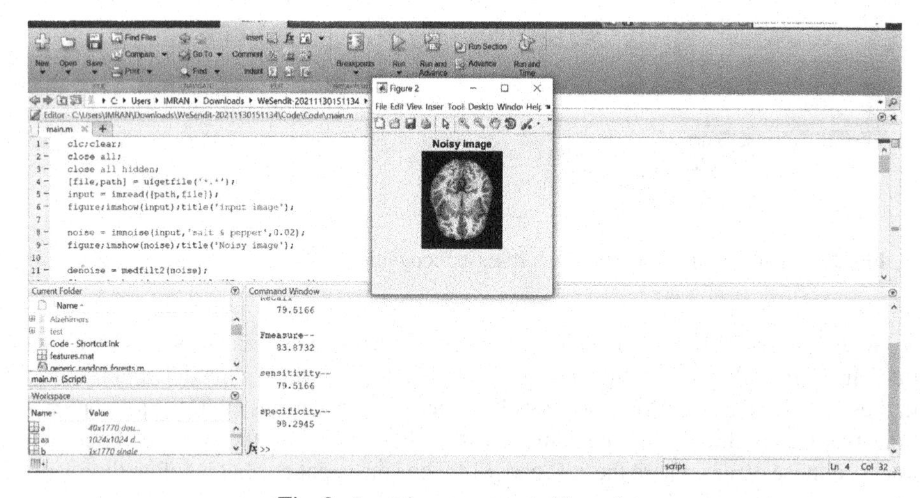

Fig. 3. Input image corrupted by noise.

Figure 3 shows the noisy input image and Fig. 4 shows the denoised image using proposed method. Training accuracy of the proposed method shown in Fig. 5. Table 1 shows the comparison of training accuracy of the proposed method with various defined methods. An accuracy comparison of proposed method with various defined methods was shown by bar graph in Fig. 6. It shows the proposed has descent advantage of training accuracy compared to existing methods. The classifications using the earlier models are having lower accuracy rates. The accuracy in detection in MCI vs CN using Xception model is 89.23%, MobileNet is 89.89%, ensemble learning + CNN is 79%, Hypergraph + Support Vector machine is 69.14% and the proposed method achieves

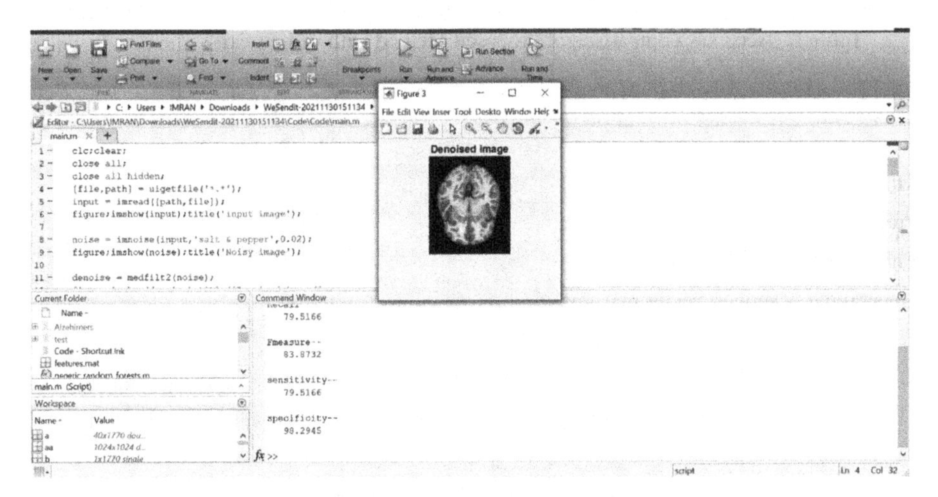

Fig. 4. Denoised image

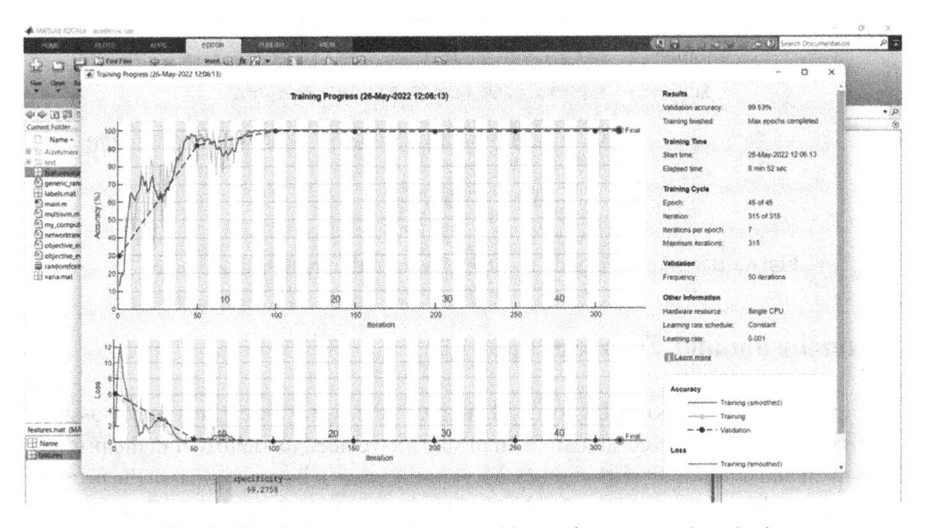

Fig. 5. Graph showing accuracy and loss using proposed method

99.53%. Accuracy comparison shows that the proposed model is perform better than the various defined methods for classification of MCI vs CN.

Table 1. Accuracy comparison of Proposed Method with various existing methods

Model	Accuracy
Xception	89.23
MobileNet	89.89
Proposed method	**99.53**
Ensemble learning + CNN	79.00
Hypergraph + support vector machine	69.14

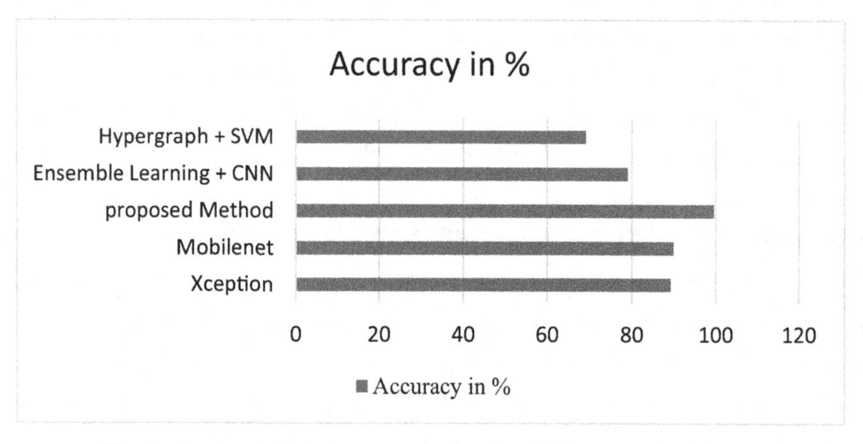

Fig. 6. Bar graph depicting accuracy levels of different methodologies

5 Conclusion and Future Scope

This paper mainly focuses on diagnoses Alzheimer's disease at the earliest stage possible. As Alzheimer's cannot be curable in its final stage's we need to identify it in the primary stages itself and diagnose it if its curable. Most of the cases when identified in the primary stages are curable so the need for this kind of projects is more useful and advantageous. All the advanced Machine learning algorithms are used to improve the accuracy of Alzheimer's detection.

The identification of MCIc and CN phases is the main focus of this investigation. Since MCIc is the initial stage of AD, grouping precision between MCIc and CN is only discussed in this work. Three previously created models—SVM, Neural Networks, and Fine Decision Tree are coupled in order to improve the grouping precision of MCIc vs CN. Results demonstrate that the gathering model provides improved order precision by utilizing the separate convolution layer of both SVM and neural networks. The method offers a positive result in terms of identifying AD's early stages. To carry out the early detection of AD, grouping precision of MCIc vs MCInc needs to be increased. This task could be extended further.

References

1. Dileep, M., Prudhvi Raj, B., Prasannakumar, G.: A new image encryption and data hiding technique using wavelet transform. Int. J. Res. Eng. Technol. **5**(5), 195–198 (2016)
2. Prudhvi Raj, B., Prasannakumar, G., Elisha Raju, B.: Hiding an image in an audio file using LSB audio technique. In: International Conference on Computer Communication and Informatics (ICCCI), pp. 1–4 (2021)
3. Ramesh Chandra, K., Prudhvi Raj, B., Prasannakumar, G.: An efficient image encryption using chaos theory. In: International Conference on Intelligent Computing and Control Systems (ICCS), pp. 1506–1510 (2019)
4. Thapa, S., Singh, P., Jain, D.K., Bharill, N., Gupta, A., Prasad, M.: Data-driven approach based on feature selection technique for early diagnosis of Alzheimer's disease. In: International Joint Conference on Neural Networks (IJCNN), pp. 1–8 (2020)
5. Francis, A., Pandian, I.A.: Early detection of Alzheimer's disease using ensemble of pre-trained models. In: International Conference on Artificial Intelligence and Smart Systems (ICAIS), pp. 692–696, (2021)
6. Salehi, A.W., Baglat, P., Sharma, B.B., Gupta, G., Upadhya, A.: A CNN Model: earlier diagnosis and classification of Alzheimer disease using MRI. In: International Conference on Smart Electronics and Communication (ICOSEC), pp. 156–161 (2020)
7. Neelaveni, J., Devasana, M.S.G.: Alzheimer disease prediction using machine learning algorithms. In: 6th International Conference on Advanced Computing and Communication Systems (ICACCS), pp. 101–104 (2020)
8. Jiménez-Mesa, C., et al.: Optimized one vs one approach in multiclass classification for early Alzheimer's disease and mild cognitive impairment diagnosis. IEEE Access **8**, 96981–96993 (2020)
9. Lodha, P., Talele, A., Degaonkar, K.: Diagnosis of Alzheimer's disease using machine learning. In: Fourth International Conference on Computing Communication Control and Automation (ICCUBEA), pp. 1–4. (2018)
10. Khan, R U., Tanveer, M., Ram Bilas, P.: A novel method for the classification of Alzheimer's disease from normal controls using magnetic resonance imaging. Expert Syst. **38**(1), (2020)
11. Zhu, Y., Kim, M., Zhu, X., Kaufer, D., Guorong, W.: Long range early diagnosis of Alzheimer's disease using longitudinal MR imaging data. Med. Image Anal. **67**, 101825 (2021)
12. Sivakani, R., Ansari, G.A.: Machine learning framework for implementing Alzheimer's disease. In: International Conference on Communication and Signal Processing (ICCSP), pp. 0588–0592 (2020)
13. Ruyi, X., et al.: Early diagnosis model of Alzheimer's disease based on sparse logistic regression with the generalized elastic net. Biomed. Signal Process. Control **66**, 102362 (2021)
14. Chitradevi, D., Prabha, S.: Analysis of brain sub regions using optimization techniques and deep learning method in Alzheimer disease. Appl. Soft Comput. **86**, 105857 (2020)
15. Xiao, R., Cui, X., Qiao, H., et al.: Early diagnosis model of Alzheimer's disease based on sparse logistic regression. Multimed. Tools Appl. **80**, 3969–3980 (2021)

References

Cyber Security and Networking

Publishing Data Objects in Data Aware Networking

G. K. Mohan Devarakonda[1](✉), I. Kali Pradeep[1], M. Venkata Durga Rao[1], and Y. K. Sundara Krishna[2]

[1] Vishnu Institute of Technology, Bhimavaram, India
[2] Department of Computer Science, Krishna University, Machilipatnam, India
krishnamohan.dg@vishnu.edu.in

Abstract. In the era of digital world everyday huge amount data is generating in the form of audio, video, image, spread sheets, documents etc. Handling such huge data with existing conventional networks is critical and a major issue. International Telecommunication Union–Technology (ITU-T) identified relevant issues and proposed characteristics and capabilities of future networks. One such system is Data Aware Networking which ought to provide effective and efficient information system to the users with most updated data. ITU-T stated that data object is a unit of information in Data Aware Networking. Data objects are identified either by a unique ID. The information sharing is done by the way of publishing Named Data Objects (NDO) and delivering to the clients [10]. In the context of DAN, the unit of information will be called as a Data Object. In order to provide authentic and most updated data, there is a need to maintain DAN data in a database system. This paper proposes a structured model of storing the NDOs in a relational model for efficient maintenance of data objects and for effective routing of NDOs between clients, publishers and DAN agents.

Keywords: Data aware networking · Named data objects · Data driven networking

1 Introduction to Data Aware Networking

Information systems play a key role in most of the contextual and digital systems now a days [1]. World Wide Web generates huge amount of data from several domains. All such information available on web domains is accessible to the public. Consistency, correctness and legality are the some of the issues with this information. In a conventional web domain several persons may involve in hosting the data. The responsibility and credibility of such persons are lacking in these systems. Also, the maintenance of correlation between information, collecting consolidated information is difficult.

N. Gupta et al. (Eds.): IC4S 2022, LNICST 472, pp. 69–75, 2023.
https://doi.org/10.1007/978-3-031-28975-0_6

ITU-T identified such issues and proposed the concept of Data Aware Networking [6, 8] to handle huge amount of data. Subsequently ITU also released its recommendation to have Data Aware Networking (DAN) with data object as primary data unit [2, 3]. The proposed architecture of DAN is shown in Fig. 1.

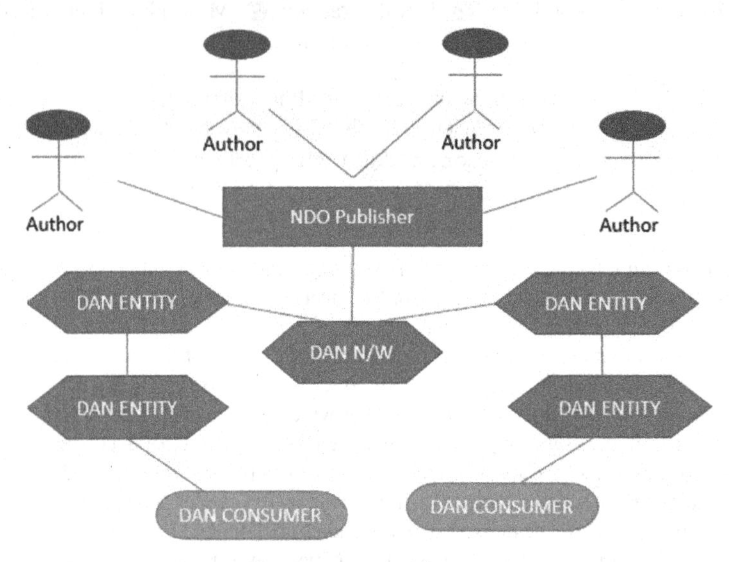

Fig. 1. DAN Architecture

The major functional units of DAN are NDO Publisher, NDO Client and DAN Agents. Every data object will be allotted a unique ID and name. Such named entities are called as Named Data Objects (NDOs). Any author or person willing to publish information, such information shall be sending to the authorized NDO Publisher of DAN. The NDO Publisher get all other required information from the author and generate NDO with name and ID. The generated NDO contains a header and actual information. The details of such Named Data Objects updated to all the Data Agents in the DAN.

NDO publisher gather required details of data item to be published from the author and build NDO and Publish the NDO. The steps in NDO publishing are preparing of header of NDO, building of NDO and publishing of NDO. The Sect. 2 describes entities of DAN system. Section 3 describes about header of NDO. Building of NDO explained in Sect. 4 and Sect. 5 describes about publication of NDO.

2 Major Components in DAN System

DAN system is networking of authors, publishers, agents and clients. Authors have to register themselves with available list of publishers to host their information. The author shall submit all personal and official details for registration. The publishers are responsible to publish information of author in a systematic way. Publishers also maintain a catalogue of published articles by different authors. Agents are middle layer in the

DAN system. Agents maintain catalogue of articles published by almost all authors and respective publishers. Agents also maintain a partial bank of articles which may be prominent articles to the nearby clients. Clients are end users of articles. This section briefly describes entities of the DAN system.

1. **Author**

Author is the primary source of information. Author may be an individual person, a group of persons, society, organization, Government or any author authority authorized to release information to the public.

2. **Named Data Object (NDO)**

NDOs are the actual data objects published by the author through publisher. A Data object is any information either for public or private purpose. Data Objects are maintained as files in database. Information to be published is encrypted and attached to document header for publication of data objects. These data objects will acquire a unique styled name and id before they are published for use.

3. **NDO Publisher**

NDO Publisher will hold the responsibility to publish the articles of the author. NDO Publisher is a Server which is full pledged computing and database system in DAN. NDO Publisher is responsible to register authors, verifying articles of author, maintaining of digital catalogue of published items, preparation header for new articles of author and preparing of NDO.

4. **NDO Client**

NDO Client is actual user of data objects in DAN system. NDO Client may be any electronic gadget attached to the DAN as leaf item in the topology DAN system. These gadgets are portable computing systems capable sending and receiving information to and from DAN network.

5. **DAN Agent**

DAN Agent is an intermediate node in the DAN system. DAN Agents acts as router for NDOs to route NDOs from NDO Publishers to NDO Clients. DAN Agents can also route requests of DAN Clients to the concerned NDO Publisher. DAN Agents also acts as mini servers to provide NDOs to NDO clients. DAN Agent maintains a database of prominent probabilistic NDOs of various Authors.

3 Header of a Named Data Object

Header is preamble of Data Object. Each data object is attached with header containing the details of Data Object. The attributes of header are Header_Size,

Data_Item_Size, NDO_Name, NDO_Id, Category, Country, Author_Id, Publisher_Id, Time_of_publishing, Volume_No, Journal_No, Data_format, Security_code, Confidentiality_code, Legal_rights_code, Price, Currency, Agent_group_code, Life_time, Update_Frequency and Reserved_bytes. Author will send all the values of these attributes to the publisher along with data item for preparation of NDO. Header itself is enough to identify a data object uniquely and ID is generated in such a way that itself is self-descriptive of the data object. Figure 1 shows structure of header of NDO. Header is prepared using UTF-8-character set.

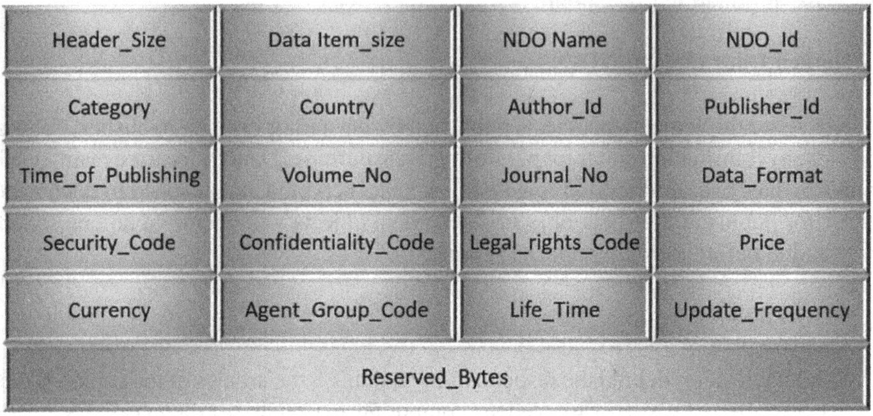

Header_Size	Data Item_size	NDO Name	NDO_Id
Category	Country	Author_Id	Publisher_Id
Time_of_Publishing	Volume_No	Journal_No	Data_Format
Security_Code	Confidentiality_Code	Legal_rights_Code	Price
Currency	Agent_Group_Code	Life_Time	Update_Frequency
Reserved_Bytes			

Fig. 2. Proposed Structure of NDO Header

The attributes of Structure of NDO are uses UTF- Character Set. The size of attributes of header NDO is given in Fig. 2. Except *Category* and *Data format* all other attributes are self-explanatory. *Category* represents type of data object. There are several categories of data objects like word, pdf, image, spreadsheet etc…., will be published viz., Government orders, statutory documents, legal documents, centuries documents, news articles, entertainment articles, text books, novels, and so on. One can address in this issue. Each data item will be identified by using a unique Item_id irrespective of the category of the object.

4 Building a Named Data Object

Named data objects are the main entities in Data aware networking. Maintenance of these objects is a challenging issue [5]. The copies of published Data items (NDOs) are cached in several Data Agents. Keeping consistence, correctness and versions are some of the related issues. One can address these issues by maintenance of proper structure of header of NDO, building of NDO, and proper database systems at NDO Publisher and Data Agents.

The actual information is converted in encrypted form and header is prepared as per the structural of header as shown in Fig. 2. The encrypted data item is in binary form is attached to the header [4]. The structure NDO is shown in Fig. 3.

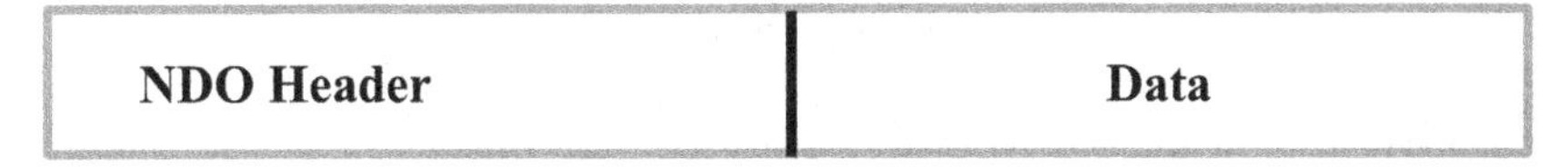

Fig. 3. Proposed Final Structure of NDO

5 Publishing a Named Data Object

The NDO Publisher builds NDOs and publishes in DAN system. Every NDO Publisher maintain a full pledged database containing the details all self-published NDOs, details of all DAN Agents in DAN system, each publication wise prominent probabilistic DAN agents, particulars other NDO Publishers, details all registered authors, details categories of data objects, publication policies, legal constraints, particulars of backend database servers etc. (Table 1).

Table 1. The size of attributes of header NDO

Description of fields of header of NDO (UTF-8 character set)		
S No	Field	No of bytes
1	*Header_Size*	4
2	Data_Item_Size	4
3	*NDO_Name*	200
4	NDO_Id	4
5	*Category*	4
6	*Country*	4
7	*Author_ID*	4
8	*Publisher_ID*	4
9	*Time_of_publishing*	78
10	*Volume_No*	4
11	*Journal_No*	4
12	*Data_format*	4
13	*Security_limits_code*	4
14	*Confidentiality_code*	4
15	*Legal_rights_code*	4
16	*Price*	4
17	*Currency*	4
18	*Agent_group_code*	4
19	*Life_time*	78

(*continued*)

Table 1. (*continued*)

Description of fields of header of NDO (UTF-8 character set)		
S No	Field	No of bytes
20	*Update_Frequency*	78
21	*Reserved_bytes*	118
Total size		600

The database of NDO publisher is updated immediately after an update is available from the author which will be update with all access points of DAN Network.

6 Conclusion

Named Data Objects (NDOs) are primary data units in DAN system. Building NDO is significant stage in success of DAN system. Critical items in building of NDO are encrypting of original data object of author, preparation of header and attaching encrypted data object to header. As demand of century's information is high, such data objects also to be included in publisher's database.

One can design DAN system on top of existing conventional web domain client server system hiding the details. Conventional routers are different from DAN Agents (DAN Routers). DAN Agents are purely software routers build on top conventional network. One can address issues such as security issues that arise a part of integrating DAN with conventional network layers [7].

References

1. Yu, K., Eum, S., Kurita, T.: Information-centric networking: research and standardization status. IEEE Access **7** (2019)
2. Koteswara Rao, O., Sundara Krishna, Y.K., Mohan Devarakonda, G.K.: Routing algorithms for data driven networking. Int. J. Comp. Sci. Eng. **7**(4) (2019)
3. Koteswara Rao, O., Sundara Krishna, Y.K., Mohan Devarakonda, G.K.: Reactive routing algorithms for data object routing in data aware networking. Int. J. Res. Electr. Comp. Eng. **7**(4) (2019)
4. Ravi Kumar, M., Sujatha Lakshmi, V., Sundara Krishna, Y.K.: A novel structure of data objects in data aware networking. Int. J. Eng. Res. Comp. Sci. Eng. **5**(4) (2018)
5. Sujatha Lakshmi, V., Sundara Krishna, Y.K.: Contextual framework of data object in data aware networking. Int. J. Appl. Eng. Res. **13**(23) (2018)
6. ITU-T Y.3071 Telecommunication Standardization Sector of ITU Series Y: Global Information Infrastructure, Internet Protocol Aspects, Next-Generation Networks, Internet Of Things And Smart Cities Future networks Data aware networking (information centric networking) – Requirements and capabilities (2017)
7. Kreutz, D., Ramos, F.M.V., Esteves Verismo, P., Esteve Rothenberg, C., Azodolmolky, S., Uhlig, S.: Software-defined networking: a comprehensive survey. Proc. IEEE **103**(1), 14–76 (2015). https://doi.org/10.1109/jproc.2014.2371999

8. ITU-T Y.3033 Telecommunication Standardization Sector of ITU Series Y: Global Information Infrastructure, Internet Protocol Aspects and Next-Generation Networks – Future networks Framework of data aware networking for future networks (2014)
9. ITU-T Y.3001-Telecommunication Standardization Sector of ITU Series Y: Global Information Infrastructure, Internet Protocol Aspects and Next-Generation Networks – Next Generation Networks – Future networks Future networks: Objectives and design goal (2011)
10. Lopez, J.E., Arifuzzaman, M, Li, Z.: Seamless mobility in data aware networking. In: ITU Kaleidoscope Academic Conferences (2015)

YouTube Comment Analysis Using Lexicon Based Techniques

Mohan Sai Dinesh Boddapati, Madhavi Sai Chatradi, Sridevi Bonthu(✉), and Abhinav Dayal

Vishnu Institute of Technology, Bhimavaram, Andhra Pradesh, India
{20PA1A0521,20PA1A0526,sridevi.b,abhinav.dayal}@vishnu.edu.in

Abstract. YouTube is used to watch music videos, comedy shows, how-to guides, recipes, hacks, and more. As of February 2020, more than 500 h of video were uploaded to YouTube every minute. This equates to approximately 30,000 h of newly uploaded content per hour. The amount of content on YouTube has increased dramatically as consumers' appetites for online video have grown. Indeed, the number of video hours uploaded every 60 s increased by roughly 40% between 2014 and 2020. The direct means of user review for this content is the comment section. To survive this cut-through competition, the content creators should constantly check up on their viewers' opinions, their reviews, and their sentiments toward the video. Although comments provide a direct means of feedback, the YouTuber cannot actually read all those comments. There may be times when he wants to know the drawbacks of the video. This paper proposes a dashboard that assists the content creators in actually looking at the positivity and negativity they have gained through the video. We opted for lexicon-based techniques over traditional classification to classify the comments into various categories. This project is a boon to the content creator's ability to increase his viewership.

Keywords: Summarization · Comment classification · Natural language processing · Polarity · YouTube comments

1 Introduction

1.1 Content Creation

With 2 billion monthly active users, YouTube is the second most popular social media platform based on total visitors and page views. It has gained huge popularity among content creators. A large number of content creators upload their video content on this platform. These videos get tonnes of views and comments. Content creators need to continuously work on maintaining the quality and quantity of their content. To do so, they must collect feedback from their viewers. This feedback lets them understand the influence of their creations. In addition to improving audience engagement, feedback also provides information on the aspects of the content that need improvement. The Youtube comment section is a direct means of feedback from the users.

N. Gupta et al. (Eds.): IC4S 2022, LNICST 472, pp. 76–85, 2023.
https://doi.org/10.1007/978-3-031-28975-0_7

1.2 YouTube Revolution

YouTube comments are an opportunity for the site's 1.7 billion unique monthly visitors to share what they love, hate, or simply must troll. Comments can also be a powerful opportunity for positive community building and, at the same time, can be a place for negativity.

It is part of many people's social media strategy to convince the audience that they want to make the most of their presence. Managing comments effectively (with moderation, replies, and analysis) is critical. The comment section booms within seconds for most of the popular channels, and it is a difficult job for the content creator to manually read all those comments. We created a dashboard where the content creators get a glance at the positivity and negativity of their content.

1.3 Contributions

The main contributions of this work are

- The classification of raw comments into five categories, viz., highly-positive, positive, neutral, negative, and highly-negative comments, and later summarises them as positive, negative, and neutral comments. This approach helps the content creators to increase their viewership and analyse the opinions of the content viewers.
- Analyze the video based on computing rank by considering the intensity of positive and negativeness in those raw comments. So that it helps the viewers decide to watch the video or not based on the rating computed.
- Creation of a user-friendly dashboard that helps YouTubers to know about information related to that video, classified comments, summarization of comments, and the rank of that video.

YouTube videos could be a powerful medium to spread information because the audience can access them easily. Also, it could describe the public opinion that could affect national unity. YouTube helps in various sectors like education(E-learning), entertainment, etc.

2 Related Work

There has been much sentiment analysis on YouTube. These works involve collecting and analysing YouTube comments from various areas in order to gain significant and interesting insights.Currently, YouTube has over 122 million active users on a daily basis. 1 billion hours of content are watched across the world every day 62.

There are many ways to classify raw comments: machine learning models and NLP-based algorithms are used for sentiment analysis.

2.1 Machine Learning Models

Many researchers build machine learning models for the classification of raw comments by using supervised and unsupervised techniques. Singh et al. built a classification model for classifying raw comments by using SVM, Naive Bayes, and KNN algorithms [1]. They analysed their model with various algorithms for better classification of comments and for good accuracy. Hajar et al. [2] built a text-based emotion detection system, based on an unsupervised machine learning algorithm by using YouTube comments as a data corpus. They achieved an average precision of 92.75%, and a 68.82% average accuracy by using SVM as a machine learning algorithm. Wadhwani et al. [3] built a machine learning model for sentiment analysis of YouTube comments. They achieved an accuracy of 81% but proposed an accuracy of 89.3% by using performance metrics.

2.2 NLP Based Algorithms

Nowadays, NLP is a growing technology, frequently used for sentiment analysis, language modeling, speech recognition, etc. Many researchers are using NLP for sentiment analysis. Asghar et al. [4] classified raw comments based on polarity-based techniques. They used several categories of videos, like news, comedy, shows, etc. They evaluated their model by using performance metrics and used SVM for classification. Olga Uryupina et al. [5] classified comments based on a corpus containing labels. It is an annotation project for text categorization and targeted opinion mining of user-generated comments on YouTube videos. Potthast et al. [6] worked on summarizing and visualizing the opinions expressed in the form of Web comments. Poche et al. [7] worked on Analyzing user comments on YouTube coding tutorial videos and summarizing those comments by using a dataset of 6000 comments and building an SVM model with an average accuracy of 77%. Kalra et al. [8] worked on YouTube comments by scraping the YouTube comments using Selenium, requests, and Beautiful Soup and using various evaluation metrics for Random Forest Classifiers. We analyzed 300 videos collected from various sectors to classify comments, video ratings, and topic-based recommended videos. Many researchers show various ways to analyze the comment classification of raw comments by building prediction models on top of them. We used NLP-based techniques to build a model.

3 Data Collection

The required data is collected by using the following two techniques.

- YouTube API and
- Webscraping

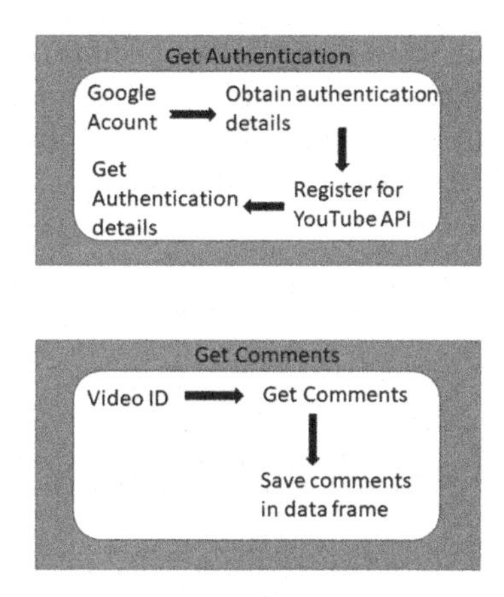

Fig. 1. Working of YouTube API

3.1 YouTube API

Extracting information from the web is pretty cool, and it becomes more professional when you use APIs provided by an application. YouTube is the most popular video-sharing platform, and it is owned by the world's biggest tech giant, Google. So, extracting comments would be easier. There are several ways to extract comments from YouTube. Firstly, set up the YouTube API and get the credentials to obtain the API key. With that key, we can extract comments by using YouTube ID. The process of working of YouTube API is clearly described in Fig. 1. The major drawback is that we can only extract 100 comments by using this technique.

3.2 Web Scrapping

3.3 YouTube API

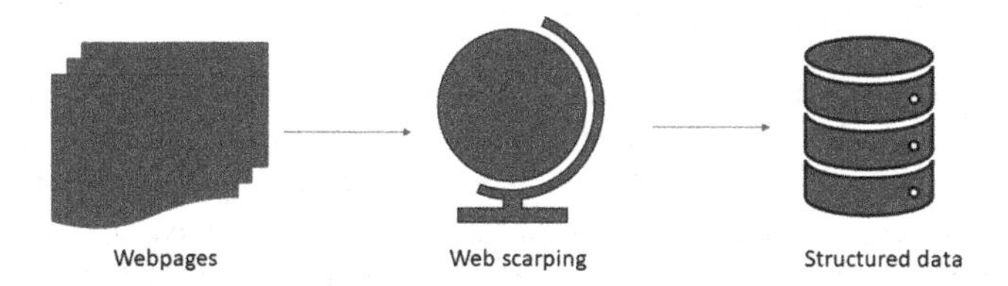

Fig. 2. Outline of web scrapping procedure

To overcome the problem faced during the extraction of comments using the YouTube API, we used the scraping technique. As shown in Fig. 2, web scrapping algorithms take web pages as input and produces structured output. Generally, the Python Selenium[1] and BeautifulSoup[2] libraries are used to scrape the comments manually by scrolling down the pages. Although extracting data from YouTube can be done with the help of web scraping, it is slow, and most of the big web applications do not promote web scraping.

4 Methodology

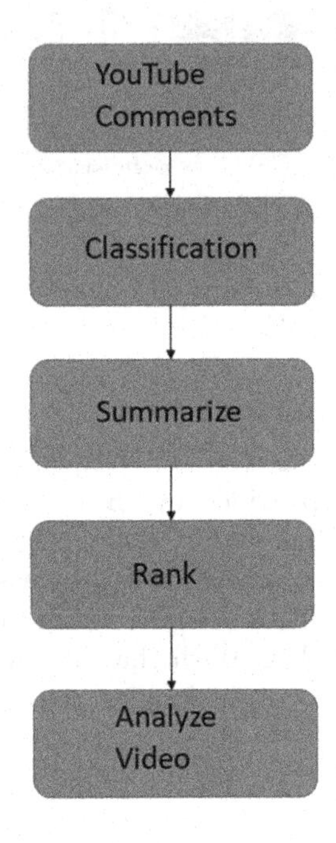

Fig. 3. Architecture of the proposed work.

The architecture of the proposed work is shown in Fig. 3.

4.1 Extraction of Comments

We extracted raw comments by using the YouTube API and scraping techniques. Fig. 1 focuses on the extraction of raw comments from the YouTube API by authorising the account and registering for the YouTube key to get those raw comments. The main drawback of this technique is that we can extract only 100 raw comments based on either relevance or the newest comments.

To overcome that problem, we used the scraping technique by using Selenium and BeautifulSoup libraries in Python so that the comments are extracted manually and stored in a data frame. The drawback of this technique is that it takes more time to scrap all those raw comments.

[1] https://www.selenium.dev/.
[2] https://pypi.org/project/beautifulsoup4/.

4.2 Classification

Those raw comments are further classified into five categories. Let the training data $D = \{X, y\}$ where X is a set of raw comments, y is a class label $\in$ {Highly-Positive, Positive, Neutral, Negative, and Highly-Negative} for each raw comment, x_i in X that maps to the corresponding class label in y. A sample set of training examples are shown in the Table 1.

Table 1. Sample Training examples after scrapping.

Comment	Label
Mam, your presentation is too excellently good and your smile is too really good mam .....!!!!	Highly positive
It's a good and nice video to watch	Positive
Ma'am please make a fast video I req please daily Make 3 videos per day	Neutral
Mind blowing eyes	Positive
Ur are saying something wrong in java course	Negative
Tq mam for your kind-hearted patience and we love you so much for your excellent teaching skills your way to speaking helps a lot in campus placements	Highly positive
You are looking nice mam and your speaking is good	Positive
Whatever your teaching is not understood by me and your way of speaking is too worst and don't make any videos right now ?	Highly negative
It's really a wrong question at 20:03 s don't say wrong to students	Negative

There are different approaches to classifying the sentiment of the text. We used lexicon-based modelling in NLP for the classification of raw comments. With lexicon-based sentiment analysis, words in texts are labelled as positive or negative (and sometimes as neutral) with the help of a so-called valence dictionary. Consider the phrase, "Good people sometimes have bad days." The word "good" is labelled as positive, the word "bad" as negative, and possibly the other words as neutral. The TextBlob[3] library in Python helps to compute the polarity of a sentence or word that determines the sentiment of that sentence or word and helps in the classification of raw comments. The polarity ranges from $[-1, 1]$, where -1 indicates negative polarity, 1 indicates positive polarity, and 0 indicates neutral polarity. The polarity values for a given set of sample sentences is tabulated in Table 2. Based on a range of polarity, we classified those raw comments.

[3] https://textblob.readthedocs.io/en/dev/.

Table 2. Polarity values of some sentences.

Sentence	Polarity
I love Monday but I hate Monday	−0.15
Miss Universe is beautiful	0.85
I love California	0.5
The covid pandemic is really terrific	−0.9
I'm travelling	0.0
We are having fun there	0.3
We hate you	−0.8

4.3 Summarization

Let X be a set of classified comments *for eg. positive or negative or neutral).* This work used Sumy library of python for doing summarization. Now X can be summarized into W where ($W << X$) so that it helps the YouTubers to look at the overview of comments.

4.4 Dashboard

Creating a user-friendly dashboard helps both content creators and users. For content creators, it helps to look at the view counts, classified comments, summarization of classified comments, and analyze the video. For users it helps to watch a video and analyze the video to watch or not based on rank. We have created the dashboard by using Python Streamlit[4] library.

4.5 Extensions

It's difficult to analyze the video that provides users with the best content based on the YouTube recommendation system, to overcome this problem we computed the rank for the top 5 recommended videos by YouTube so that it helps users to watch the best video based on their search as shown in Fig. 4.

4.6 Rank

To analyse the video's best or not views and likes count doesn't play a major role because for a channel having more number of subscribers will easily get more views and likes therefore based on intensity of comments it's easy to analyse video. So to analyse video by computing rank determines the video is best or not. Rank Algorithm helps to compute rank by considering the emotional intensity of positive and negative comments along with their count.Based on rank value a star rating is given to each video that helps for YouTube users (Figs. 5, 6, 7, 8 and 9).

[4] https://streamlit.io/.

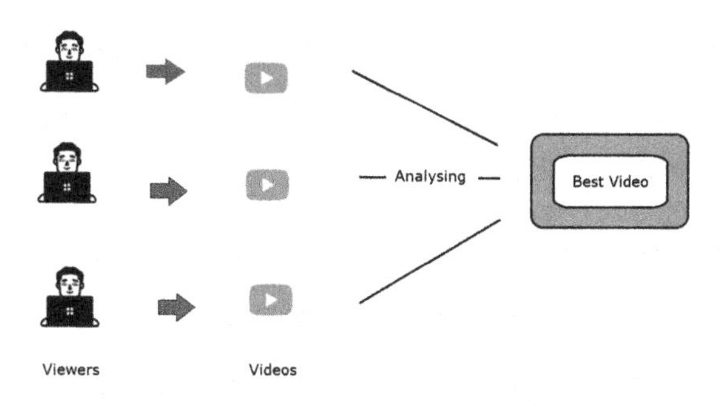

Fig. 4. Recommending the best video

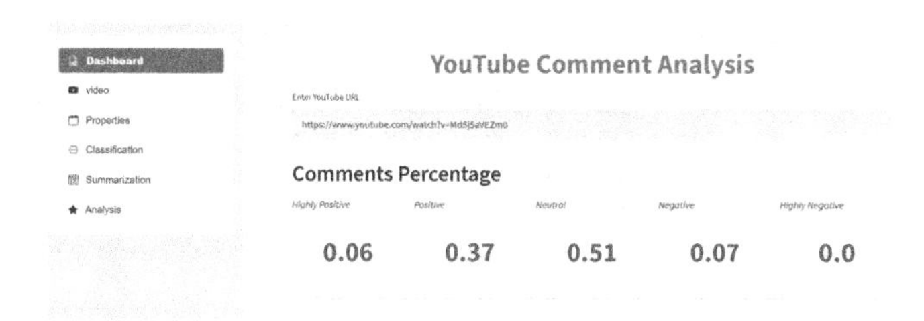

Fig. 5. Home screen of the dashboard

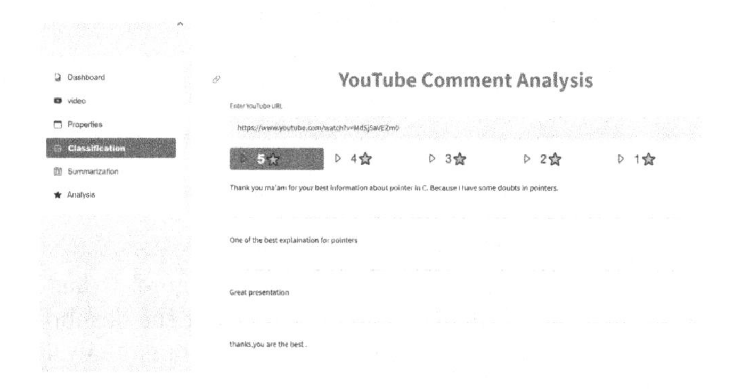

Fig. 6. Sentiment Classification of the video

Fig. 7. Summary of the video

Fig. 8. Rank analysis

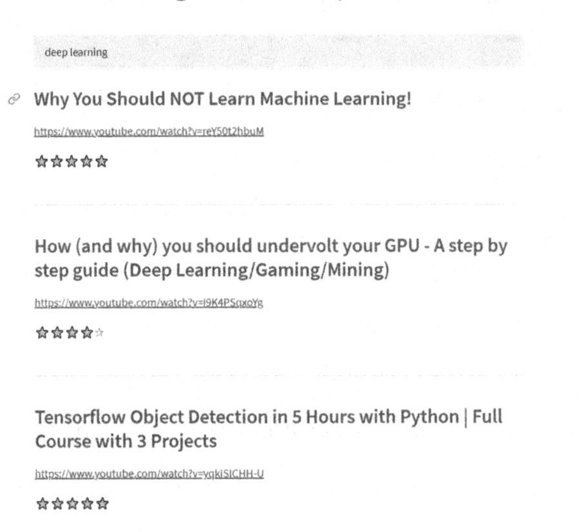

Fig. 9. Prediction of the best videos

5 Results

This section presents the dashboard snapshots.

The homescreen of the dashboard is shown in the figure. It has provision to enter the URL of the respective video. The left panel of the dashboard contains the menu leading to classification, summarization, and rank analysis. The dashboard displays the probability of all five classes for the comments of the entered YouTube video. The classification tab is displayed in the figure. It displays the count of comments belonging to every category. The figure is the summarization window, which displays the summary of the comments in positive, negative, and neutral sections. The assigned rank can be viewed in the Analysis tab as shown in the figure. Finally, the figure shows the application page, which recommends the best videos.

6 Conclusion

With the dashboard, the content creator will know the sentiment of the viewer at all times. They can have a gist of what's working and what's not working.

For instance, the dashboard could show them the positivity and the negativity of a newly released video. A brief summary of positivity and negativity increases their productivity compared to manually filtering their comments. This would save them a lot of time. There have been many classification models for sentiment analysis, and most of them were built on traditional classification techniques. We used lexicon-based techniques, which use a valence dictionary to predict the positivity of a given word and fine tune on threshold values to categorize the comments into some categories.

The comments on YouTube are a massive amount of unstructured data. Most of the existing machine learning models and algorithms can produce meaningful insights when provided with structured data. The categorization of raw comments into the above-mentioned classes adds structure to the comments, Our work can also be used to annotate these comments for building a supervised learning model or the categorized comments can be fed into some existing algorithms. In addition to these, we assigned a rank to the video based on some polarity metrics, and this can be used as a filtered search to extract the video that provides the best content.

References

1. Singh, R., Tiwari, A.: Youtube comments sentiment analysis. Int. J. Sci. Res. Eng. Manage. **5**(5), 1–11 (2021)
2. Mousannif, H.: Using YouTube comments for text-based emotion recognition. Procedia Comput. Sci. **83**, 292–299 (2016)
3. Wadhwani, S., Richhariya, P., Soni, A.: Analysis & Implementation of Sentiment Analysis of User YouTube Comments. No. 7703. EasyChair (2022)
4. Asghar, MZ., et al.: Sentiment analysis on YouTube: a brief survey. arXiv preprint arXiv:1511.09142(2015)
5. Uryupina, O., et al.: SenTube: A corpus for sentiment analysis on YouTube social media. In: Proceedings of the Ninth International Conference on Language Resources and Evaluation (LREC 2014) (2014)
6. Potthast, M., Becker, S.: Opinion summarization of web comments. In: Gurrin, C., et al. (eds.) ECIR 2010. LNCS, vol. 5993, pp. 668–669. Springer, Heidelberg (2010). https://doi.org/10.1007/978-3-642-12275-0_73
7. Poché, E., et al.: Analyzing user comments on YouTube coding tutorial videos. In: 2017 IEEE/ACM 25th International Conference on Program Comprehension (ICPC). IEEE (2017)
8. Kalra, G.S., Kathuria, R.S., Kumar, A.: YouTube video classification based on title and description text. In: 2019 International Conference on Computing, Communication, and Intelligent Systems (ICCCIS). IEEE (2019)

Information Theoretic Heuristics to Find the Minimal SOP Expression Considering Don't Care Using Binary Decision Diagrams

N. Padmavathy[1(✉)], K. S. Jhansi[1], K. Ormila[2], and B. Valarmathi[3]

[1] Department of Electronics and Communication Engineering, Vishnu Institute of Technology, Bhimavaram 08544, India
padmavathy.n@vishnu.edu.in

[2] Department of Electrical and Electronics Engineering, AMK Technological Polytechnic College, Chennai, India

[3] Department of Software and Systems Engineering, School of Information Technology and Engineering, Vellore Institute of Technology, Vellore, Tamil Nadu, India

Abstract. Logic minimization plays a significant part in decreasing the complexity of the circuit since the number of Gates will be diminished... Till today, the conventional or traditional approaches like Boolean laws; Karnaugh map; Quine-Mccluskey are in existence for Boolean expression simplification. The above approaches have several drawbacks; to name a few – logical synthesis complexity, multiple solutions using k-maps, the number of cells increases exponentially with number of variables in a Boolean function and more computational time is needed while solving using Quine Mccluskey. Moreover, the implementation of k-map and Quine-Mccluskey method is difficult in logic synthesis of chip design. The solution to all above drawbacks is the use of Binary Decision Diagrams which is faster and its applicability to large circuits is possible. The main purpose of this work is to reduce the Boolean expressions considering DC function and also to calculate the entropy of the simplified Boolean expression using binary decision diagrams.

Keywords: Binary decision diagrams · Boolean expression · Entropy · Logic gates · Karnaugh map · Minimal SOP · Quine McClusky method

1 Introduction

Over several decades, Boolean logic expression or functions had been represented as a sum-of-product (SOP) or cube form. These expressions has been transformed to simple circuits with application of Boolean algebraic rules, to achieve low logic complexity in terms of variables counts in a Boolean function (BF). There had also been other traditional, conventional approaches like k-map, tabulation method, which are being used over many decades to get minimized logic functions. No optimal solution can be achieved with use of k-Map approach because the combination of AND/OR expression varied from

N. Gupta et al. (Eds.): IC4S 2022, LNICST 472, pp. 86–111, 2023.
https://doi.org/10.1007/978-3-031-28975-0_8

person to person. In order to achieve a potential better result, the tabulation method came into existence. This method failed to compete with the Boolean rules method because of the computational complexity with increase in number of literals in a Boolean expression i.e., the minterms size is exponential to the number of input variable [1]. Motivated with the drawbacks, the authors provide a solution in minimizing the Boolean expressions with use of binary decision diagrams (BDD) and information heuristics method for calculating the entropy of the simplified BF. [2] Used BDD CAD applications (logic synthesis, formal verification) and proposed a variable ordered algorithms for micro-canonical optimization using two procedures viz., initialization and sampling.

1.1 Entropy

The idea of self-information is entropy (the data given by a random process about itself). Entropy is sufficient to study signal imitation by quiet environments. Often there are two or more different arbitrary processes. i.e., one random process represents the source of information and another represents the output of a communication medium in which the encoded source is corrupted by another random process called noise. The fundamental quantity of information theory is entropy [3, 4] and is a measure of uncertainty of an unknown or random quantity as defined using (1)

$$H(k) = -\sum_{k=0}^{k-1} p_k log_2 p \tag{1}$$

where, k is a random variable; p_k is the likelihood of occurrence of a random variable and $H\ (k)$ is the entropy

Entropy can be related to the least number of bits it would take an average to communicate information from one location to another location and is always a relative measure to a probability distribution. Entropy is measured in ***bits*** for base-2 logarithms ***Hartley*** for base-10 logarithms, ***trits*** for base-3 logarithms and ***nats*** for natural logarithms base. It may be noted that natural logarithms are usually more convenient for mathematics while the base – 2 logarithms provide more spontaneous similes. Entropy is proportional to $(-log_2 p_k)$, with the proportionality constant determining what base logarithms are taken in. Averaging over all events according to their respective probabilities, the (1) is got. It may be noted that entropy reaches a maximum iff all states are equiprobable.

1.2 Information

Information is the data that can be stored or transmitted as variables that takes different values; say binary variables can take 0 or 1 as its variable values in digital storage and in case of raw data the outcome is either biased or unbiased. Information is quantified as the number of bits it takes to symbolize a variable and is defined using (2)

$$I(k) = -\sum_{k=0}^{k-1} \log_2 \frac{1}{p_k}\ (bits) \tag{2}$$

In other words, information is connected with a result of uncertainty or un-expectance is the likelihood of the occurrence of the event. The more the unexpectedness or uncertainty of an event, higher is the information. Therefore, information is directly related to uncertainty or inversely related to the probability of occurrence of that event.

2 Literature Review

The authors of [5, 6] have proposed that the central problem of logic synthesis, Binary logic minimization, is most appropriate for reliability analysis and automated reasoning. With BDD and Disjoint Sum of Product (DSOP) reduction, the authors present a strategy for reducing the Boolean Sum of Products function [1]. Proposed on algebraic factorization method. It has been very successful and has become the most common approach in logic synthesis. However, whereas arithmetic and XOR-intensive logic functions, which can be more succinctly represented as combination, produce results that are far from satisfactory, AND/OR-intensive control and random logic functions produce results those are close to optimal. Although logic optimization techniques based on Boolean factorization may provide superior results to algebraic techniques, their high computational complexity prevented them from competing with algebraic techniques.

A new method for efficiently calculating various Shannon information measures using BDD was proposed in [7]. Algorithm for BDD reordering that outperforms other reordering methods in terms of the results it produces. The authors of this paper implemented the technique and reordering algorithm, and the results on circuit's benchmarks are analyzed [8]. Provided an exact method for utilizing BDDs to minimize logic functions. The function is mapped to an extended space using this strategy, which gives it special properties that can be used to calculate its prime and minterms. Conceptually creating a covering table whose columns represent the primes is the next step. The use of BDD as an effective method for minimizing DSOP was suggested and discussed in [9]. An advantage of using BDDs is that they implicitly represent terms. This scheme makes the algorithm faster than techniques that use explicit representation and applies it to large circuits.

Directed acyclic graphs (DAG) [10] were used to represent a BF using OBDD. Tests of functional properties like satisfiability and its equivalence are straightforward thanks to their canonical representation. On OBDD data structures, a number of BF operations can be implemented as graph algorithms. Through symbolic analysis, a wide range of issues can be resolved by utilizing OBDDs. Boolean variables are first used to encode any potential variations in the operating conditions and parameters of the system. In VLSI computer aided design, the authors of [11] emphasized that decision diagrams are the current state of the data structure and have been utilized successfully in numerous other fields. Decision diagrams are utilized extensively and are incorporated into commercial tools. A method for looking into a few families of elementary order – 2 matrices was proposed in [12]. When real vectors are used for Boolean function transformations, new transforms are introduced. One-to-one mapping in a binary/ternary vector space is what these transforms do.

A specific set of fundamental functions can be used to conceptualize arithmetic functions as series expansions in the space of complex valued functions on finite dyadic (binary relations) groups. A new algorithm transforms disjoint cubes, a simplified representation of Boolean functions, into generalized adding and arithmetic spectra [13]. Exclusive SOPs; [14] that are crucial to logic design and synthesis were proposed. An arithmetic expression that can be thought of as Reed-Muller expressions [15], for switching functions was proposed by replacing Boolean expressions with arithmetic equivalents. Most of studies have considered SOP expression minimization without DC using BDD. The overview of the literature survey indicates that BDD have extensive use in several applications that involves complexity is design. In this paper, minimization of SOP circuits has been computed with DC conditions. The methods which are used to minimizing the Boolean expression are the following methods.

- Boolean Algebra
- Karnaugh Map
- Quine Mc Cluskey

In detail, the concepts with examples on Boolean algebra, K-Map and Tabulation method is available in [16–19]. However, these methods are disadvantageous – like using K-Maps because based on combination of cell, the minimization outcomes may change. In other works, multiple minimized solutions are achievable for single logic expression. Same is the case with QC method but the simplification /minimization approach is time consuming i.e., the procedure is complex and lengthy (Ref example 1). As long as the number of variables does not exceed five or six, the map method of simplification is convenient. It becomes more difficult to determine which combinations form the minimum expression as the number of variables increases. The mapping approach makes it nearly impossible to simplify expressions for complex problems with seven, eight, or even ten variables. The fact that the process of minimization is dependent on human capabilities is another crucial point. The solution to above problems is the use of Binary decision diagrams for the reduction of the Boolean expressions. This paper leads the reader through the minimization process with help of suitable examples. The researcher has considered examples of Boolean expression with and without DCs.

Example 1: Without DC

1. Minimize the BF, f(W,X,Y,Z) = $\sum$m (0, 2, 5, 6, 7, 8, 10, 12, 13, 14, 15) using K -Maps. The solutions are $\overline{XZ} + XZ + W\overline{Z} + Y\overline{Z}$ (or) $X'Z' + YZ' + \mathrm{WXY}' + W'XZ$
2. Reduce the BF, f(W,X,Y,Z) = $\sum$m (0, 1, 3, 7, 8, 9, 11, 15) using K -Maps. The solution is $\overline{XY} + YZ$
3. Minimize the BF, f(A,B,C,D) = $\sum$m (0, 2, 5, 6, 7, 8, 10, 12, 13, 14, 15) using QC approach. The solutions are $\overline{XZ} + XZ + W\overline{Z} + Y\overline{Z}$ (or) $\overline{XZ} + XZ + \mathrm{XY} + W\overline{Z}$ (or) $\bar{X}\bar{Z} + YZ + \mathrm{WX} + Y\bar{Z}$(or) $\mathrm{X}'Z' + \mathrm{XZ} + \mathrm{WX} + \mathrm{XY}$
4. Reduce the BF, f(W, X, Y, Z) = $\sum$m (0, 1, 3, 7, 8, 9, 11, 15) using QC Method.

   The solution is $\overline{XY} + YZ$.

Example 1: With DC

1. Minimize the BF, f(W, X, Y, Z) = $\sum$m (0, 2, 5, 6, 7, 8, 10, 12, 13, 14, 15) + $\sum$d (1, 4)and using K -Maps. The solution is $W'Y' + Z' + X$
2. Reduce the BF, f (W, X, Y, Z) = $\sum$m(0, 1, 3, 7, 8, 9, 11, 15) + $\sum$d (5, 13) using K -Maps. The solution is $X'Y' + Y'Z + W'X'Z' + WXY$
3. Minimize the BF, f (W, X, Y, Z) = $\sum$m (0, 2, 5, 6, 7, 8, 10, 12, 13, 14, 15) + $\sum$d (1, 4) using QC approach. The solution is $W'Y'$
4. Minimize the Boolean expression f (W, X, Y, Z) = $\sum$m (0, 1, 3, 7, 8, 9, 11, 15) + d (5, 13) using QC Method. The solution for is $X'Y' + Z'$

2.1 Quine-Mccluskey

Quine-Mccluskey method [16–19] is popularly known as Tabulation method. In simplification of Boolean expression the adjacent minterms can be reduced. These minterms are reduced because they differ by only one literal. The minterms whose binary equivalent differs only in one place can be combined to reduce the minterms. This is the fundamental principles of the Quine-Mccluskey method. The minterms are written as their binary equivalents. A horizontal line separates each number of s categories from the number of one-syllable minterms in each group. This separation of minterms helps in searching the binary minterms that differ only in one place. Once the separation is over, each binary number is compared with every term in the next higher category, and if they differ by only one position, a check mark is placed beside each of the two terms and then the term is copied in the second column with a '-' in the position that they differ.

This process of comparison is repeated for every minterm. Once this process is completed the same process is applied to the new resultant terms which are placed in the column. These cycles are repeated until a single cycle passes without removing any more literals. The prime-implicants are the remaining terms, and all terms that did not match during the process. Summing one or more prime implicant gives the simplified Boolean expression. A minimal implicant, such as when a literal product term is removed to produce a non-implicant, is referred to as a prime implicant (PI). Prime implicant is a smallest possible product term got after the removal of all possible literal and further no more removal is possible. The product must be an implicant of a given function for it to be a PI. An essential prime implicant (EPI) is a PI that covers an output of the function that no other PI combination can cover.

2.2 Binary Decision Diagram (BDD)

The Binary Decision Diagram (BDD) is a graphical representation of Shannon's expansion of a logical function, as shown in Fig. 1.The concept of BDD [20] was first proposed and continued by [21]. A BDD is a DAG with two terminal nodes (denoted 0 or 1) of out-degree 0, and a set of variable (branch) nodes of out degree 2. Two successors to the root node are denoted by down sliding lines. In the Fig. 1, these branching nodes indicate a path for each Boolean variable value. The sink node is another name for the '0' and '1' nodes. If the LOW branch (dotted line) is taken from the root, the end point node '0' is reached, and if the HIGH branch (solid Line) is taken the end point node '1' is reached.

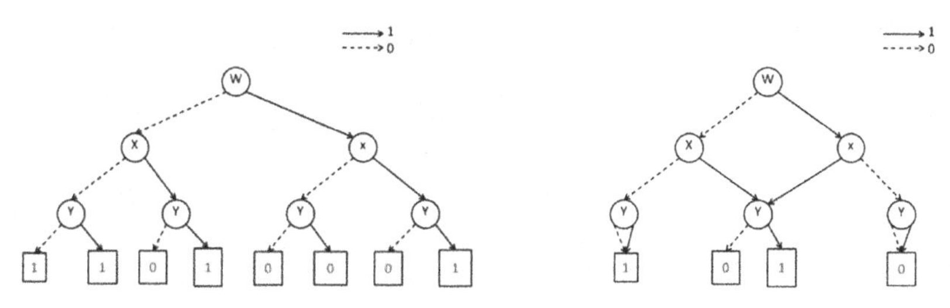

Fig. 1. Binary decision diagram **Fig. 2.** Reduced ordered BDD

Two crucial rules must be followed by the BDD: first, it must be reduced and then ordered. BDDs are well-known and frequently utilized in logic synthesis and formal verification of integrated circuits.

Ordered Binary Decision Diagram (OBDD). The OBDD extended by [22] (see Fig. 2) are based on a fixed ordering of the variables. If the variables are arranged in a linear manner across all graph paths, the BDD is ordered. OBDD ensures that the variables appear in the same order from the root to the leaves. There are not multiple occurrences of any variable along a path. As an example in the case of n-bit multiplier the number of OBDD representation of the binary multiplier increases exponentially [23]; irrespective of the variable ordering.

Recovering the significant canonicity property for a fixed variable ordering is made possible by the ROBDD representation of each BF, which is a canonical (unique) representation. By constructing their ROBDDs, we can compare BFs and determine whether or not they are equivalent. [24] Has elaborately dealt with the reducing the size of intermediate OBDDs to guarantee a most efficient OBDD in its minimized form having its milestone reference to [25]. In addition, [26] used Evolution algorithms to derive the reduced minimal paths in a BDD.

Uniqueness: Neither the LOW nor HIGH successors of any two dissimilar nodes u and v but share the same variable name. i.e.,

$$var(u) = var(v); \; low(u) = low(v); \; high(u) = high(v); \; implies \; u = v.$$

***Non redundant tests*:** There is no alike LOW and HIGH descendant for variable node u.

The application of BDD is an effective strategy for reducing DSOP. An advantage of using BDD is that terms are implicitly represented. The algorithm is faster than methods based on explicit representations, can be used on large circuits that other methods like Boolean algebra, K-map, and MC can't handle, and gets more complicated when there are more than four variables. In comparison to the methods that are currently in use, the results regarding the size of the reduced DSOPs are also superior. Fault tree analysis (FTA), Bayesian reasoning (BR), product configuration, and private information retrieval (PIR) are just a few of the lesser-known applications of BDD.

Building BDD. The BDD can be constructed for any the Boolean expression or Boolean function. Initially the decimal numbers are converted into binary equivalents and a truth table as shown in Table 1, variables are assigned to bits in the manner of their binary equivalent.

$$Let\ F(y, z) = \sum m(0, 1, 3) \tag{3}$$

Table 1. Truth table for F(y, z) = ∑m(0, 1, 3)

Decimal equivalent	y	z	F
0	0	0	1
1	0	1	1
2	1	0	0
3	1	1	1

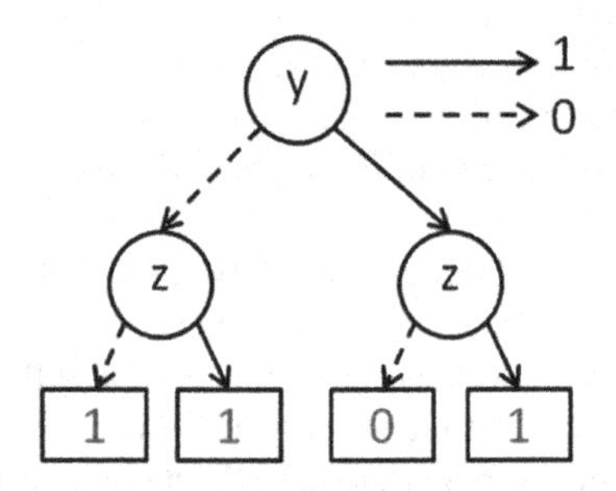

Fig. 3. Building a BDD

The first variable is taken as root *node w a*nd then if the value of the variable is 0 then it is indicated by dotted line to the next literal. If the value of the literal is 1 then it is indicated by solid line. The same procedure is followed by the next variable. In this way the BDD is constructed as shown in Fig. 3.

Reduction Rules of BDD. ROBDD can be obtained from OBDD by repeatedly applying following reduction rules. Figure 4(b) shows the reduction of the BDD using merge isomorphic rule.

Merge Isomorphic Nodes. If two nodes are having the same outputs, then, the corresponding nodes are merged. The Fig. 4(a) is the BDD representation of the BF, F (x, y, z) = $\sum m(0, 1, 3, 4, 5, 7)$. It can be understood that when root node is 'c', the terminal nodes are 1 for $F(\text{x, y, z}) = \sum m(0, 1, 3)$ with *node w* = 0 and *node x* being a DC condition (i.e., y = 0 or y = 1) as seen in subgraph1 (Fig. 4(a)). A similar type of output (subgraph2) is got for $F(\text{x, y, z}) = \sum m(4, 5, 7)$ with *node w* = 1 and DC condition (i.e.,*y* = 0 or y = 1). This implies, when $F = \sum m(0, 1, 3)$and $F = \sum m(4, 5, 7)$, same

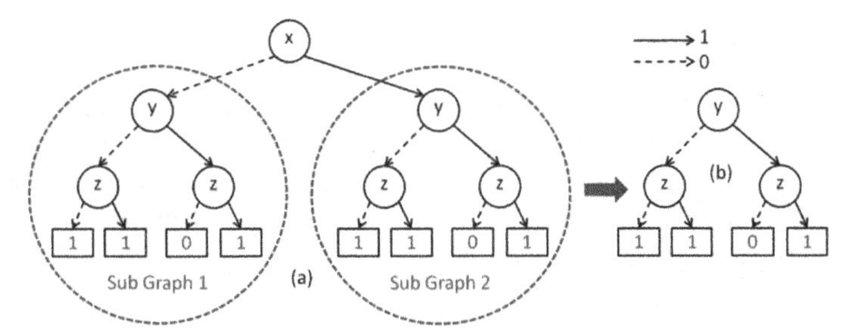

Fig. 4. (a) Isomorphic sub graphs (b) Merged Isomorphic BDD.

terminal *node a*re got. Hence, the subgraphs whose outputs are similar are termed as isomorphic subgraphs.

Isomorphic Subgraphs. The isomorphic subgraphs are the subgraphs which are similar, so the root *node w*nd one of the subgraph is eliminated. Further the elimination of similar subgraphs is done by eliminating the root *node w* and all the subgraphs, retaining one subgraph as shown in Fig. 4 (b). This process of elimination is termed as merging as seen in Fig. 4 (b).

Eliminate Redundant Tests. The two different nodes having same label and same leaves, then such nodes can be eliminated. The root node is removed if its terminal nodes are identical. This elimination is *Bottom-to-top* process. Figure 5(a) is the merged isomorphic graph. The isomorphic graph has a redundancy and can be reduced using redundant test rule. Let *node y* be the root node. According to the Fig. 5 (a), node '*y*' branches out to the terminal node 1 when $w = 0/1$. The sub graph shown within the dotted circles indicate that when *node w* is either 0 or 1, the terminal node output is 1.

Therefore, the terminal node is merged in single *node w* is seen in Fig. 5 (b). The parent (root) *node c*an be removed because the leaf (terminal) nodes are identical, and the leaf node is attached to the next higher root node (see Fig. 5(c)). The merging equivalent node means that the node which has two outputs and that are similar, then merging that node to the one output (see Fig. 5 (b)).

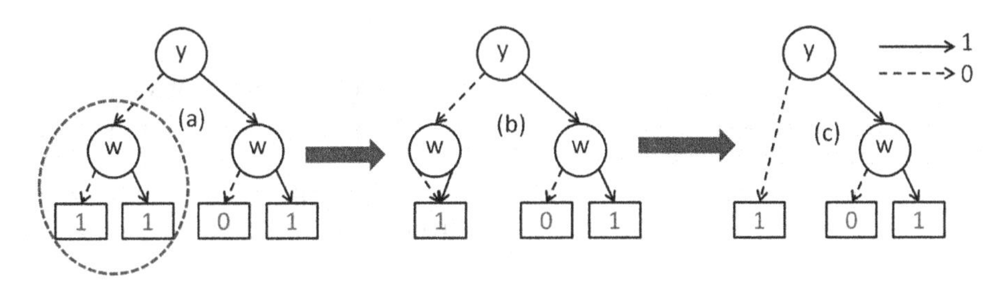

Fig. 5. (a) Merged Isomorphic graph. (b) Redundancy Elimination graph. (c) Reduced BDD

3 Methodology

The methodology involves the procedure of converting the SOP function to minimized SOP function using BDD. Based on the SOP function's variable values, the BDD is created, with the last variable marked with the function output. The SOP function is reduced, by applying the reduction rules like merging equivalent nodes and merging isomorphic sub graphs to the binary decision diagram.

4 Algorithm

The given Boolean expression generates the truth table by converting the given decimal numbers into their respective binary equivalents and obtaining their output function.

4.1 Creating BDD

The BDD can be created by the binary equivalent of a decimal number, which is present in the truth table. In every truth table, the last variable can be represented as a function output.

4.2 Reduction of BDD

The BDD is minimized by applying the reduction rules from bottom level to the top level. The reduction rules are merging equivalent leaves, merging isomorphic nodes and eliminating redundant tests. The reduction rules are thoroughly applied to the BDD to bottom level. After completion of a bottom level, reduction rules have to be applied to the next successive level. The reduction rules must be followed in this manner until the highest level is reached.

4.3 Variable Ordering Using Entropy

Step1: The variable ordering is chosen by calculating the entropy of each variable from truth table for each case i.e. by calculating the information of each node when it becomes '0' and '1'. Then, calculating the entropy by averaging the information of nodes when it is '0' and '1'.
Step 2: This process is continued until the calculation of entropy of all variables is completed. Choosing the variable as the 1stsplitting node which has the least entropy among all the variables.
Step 3: Again, calculating the entropy all nodes except 1st splitting node when a 1st splitting node is '0' and '1'. This process is continued until 'n − 1' splitting variables have been obtained.

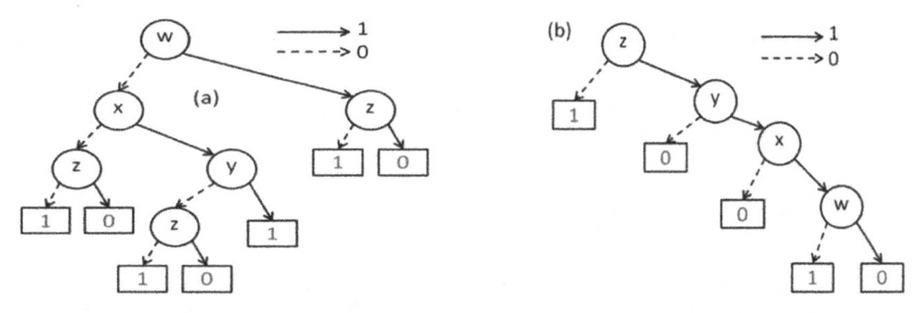

Fig. 6. (a) ROBDD with worst variable ordering. (b) ROBDD with best variable ordering

When building BDD, the right order of variables is very important. Because, as depicted in Fig. 6 (a), the number of 1-paths increases when the worst choice of variable order is chosen, which in turn increases the number of ROBDD nodes. And Fig. 6(b) shows the best variable order with less number of nodes in ROBDD.

4.4 Binate Select

Finding the DSOP from the ROBDD and finally the ROBDD equation is obtained, which is represented in covering matrix. The Binate select is done with respect to the 1st splitting variable, which is selected from entropy calculation.

4.5 Merging

The merging concept is merging the disjoint cubes from binate select. The merging process is started with the last variable of the binate select variable. This process is continued until the first variable of binate select has merged.

5 Example Considered for Study

The Boolean expression with DC has been considered to minimize the Boolean expression by applying the algorithm for the example is as shown in (4).

$$F(w, x, y, z) = \sum m(0, 1, 3, 7, 8, 9, 11, 15) + d(5, 13) \tag{4}$$

Step 1: Generation of truth table. The Table 2 is generated by converting the decimal numbers into its binary equivalents and function output is TRUE (1) if the decimal number is present in the expression. The DC conditions are marked with an "*" because merging the DC conditions is not required. If the SOP is minimized by taking into account the DC, then it is regarded as 1, and if it is not, it is regarded as 0.

Table 2. Truth table for F(w, x, y, z) = $\sum$m(0, 1, 3, 7, 8, 9, 11, 15) + d(5, 13)

Binary equivalent	W	X	Y	Z	F	Binary equivalent	W	X	Y	Z	F
0	0	0	0	0	1	8	1	0	0	0	1
1	0	0	0	1	1	9	1	0	0	1	1
2	0	0	1	0	0	10	1	0	1	0	0
3	0	0	1	1	1	11	1	0	1	1	1
4	0	1	0	0	0	12	1	1	0	0	0
5	0	1	0	1	*	13	1	1	0	1	*
6	0	1	1	0	0	14	1	1	1	0	0
7	0	1	1	1	1	15	1	1	1	1	1

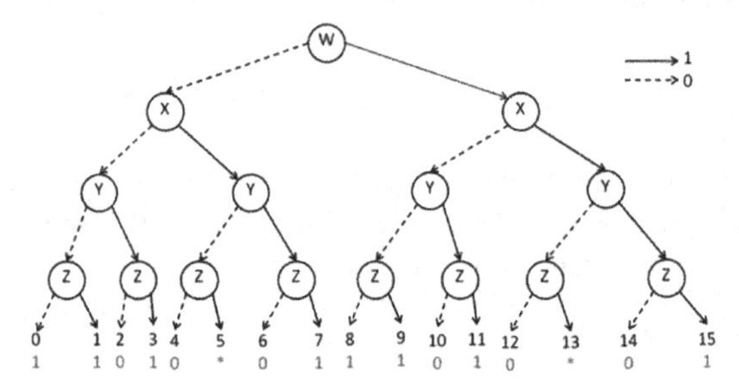

Fig. 7. BDD for **F** = $\sum \boldsymbol{m}(0, 1, 3, 7, 8, 9, 11, 15) + \boldsymbol{d}(5, 13)$.

Step 2: Creating BDD The BDD representation for the BF F = $\sum m(0, 1, 3, 7, 8, 9, 11, 15) + d(5, 13)$ is as shown in Fig. 7.
Step 3: Reducing the BDD: Figure 8, shows a BDD with the sub graphs that are similar. And so, one of the sub graphs is removed and therefore the root node is eliminated and any one of the sub graph is thus considered as depicted in Fig. 9 with the condition that node *w* is 0/1. Figure 9 depict a single x node which is got by merging the two x nodes of Fig. 8. This process is the normal reduction process.

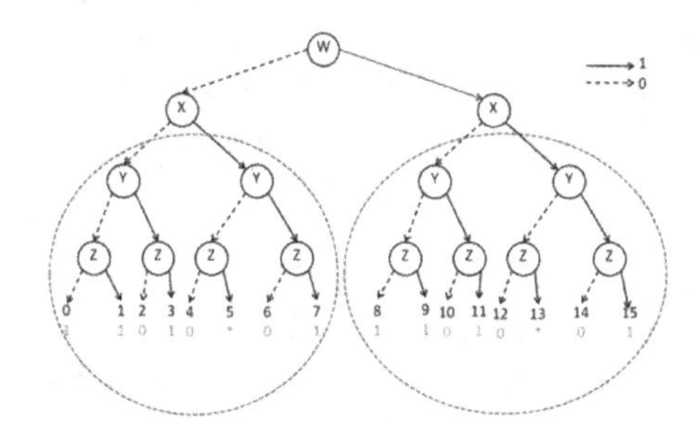

Fig. 8. Reducing the BDD by applying reduction rules

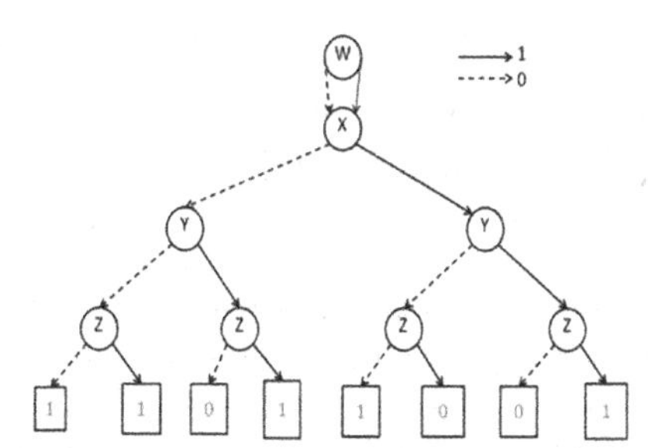

Fig. 9. Merging isomorphic sub graphs

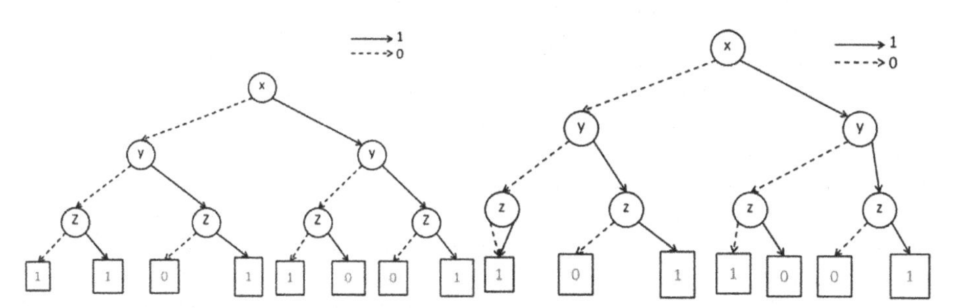

Fig. 10. After merging isomorphic nodes

Fig. 11. Merging equivalent nodes

In Fig. 9, as the outputs of *w node* are either 0 or 1, then root *node w* can be eliminated to get Fig. 10. In the left most sub graph, when the value of *y* is 0, the *node z* represents same output 1 then further *node z* is eliminated and after that the *node* y is directly assigned to 1 which results in the reduction of BDD (see Fig. 12). Figure 12, also have two identical sub graphs (see blue dotted circles of Fig. 12) when *node y* = 0. So one of them may be eliminated i.e., *node w* is reduced and its corresponding *y = 1 is* connected to *node w* (see the brown dotted circles in Fig. 13).

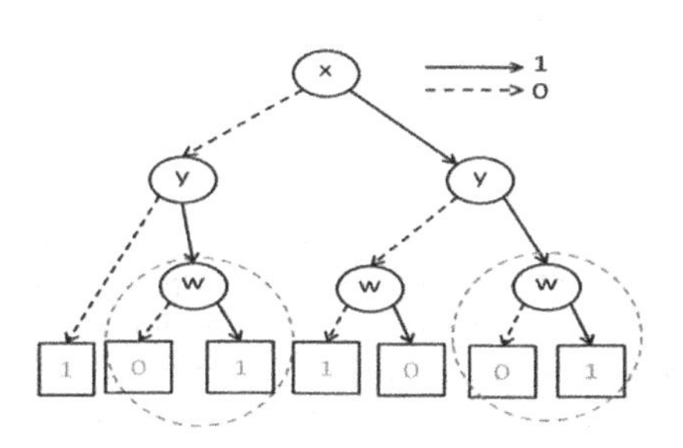

Fig. 12. After merging equivalent and isomorphic nodes in BDD

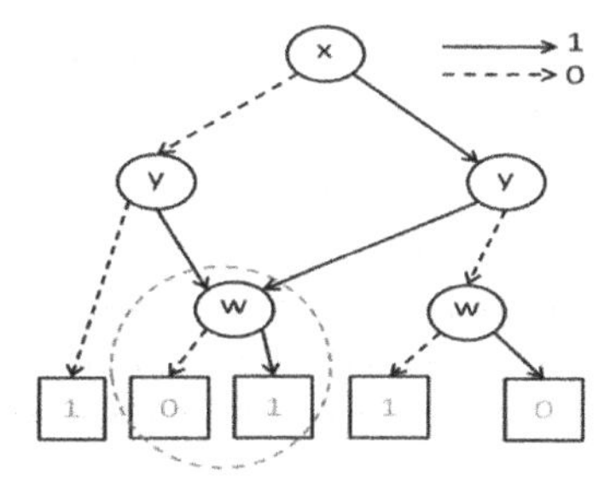

Fig. 13. ROBDD with variable BDD

The output consists of excess usage of nodes which results in increase in the nodes representation. So, therefore the excess usage can be further reduced through node elimination and merging of the isomorphic sub graphs. The output of *node x* & w is same when the value of node *y* is 1. Thus Fig. 13 is the final ROBDD.

Variable ordering by calculating Entropy by considering the DCs:
The information of each node, as well as the times when it reaches 0 and 1, can be used to calculate entropy. And averaging the values for both the cases. In this manner all the variables entropy is calculated.

$$\mathrm{I(w,0)} = (3,5) = 0.954$$
$$\mathrm{I(w,1)} = (3,5) = 0.954 \quad \mathbf{E(w) = 0.954I}$$
$$\mathrm{(x,0)} = (2,6) = 0.811$$
$$\mathrm{I(x,1)} = (4,4) = 1 \quad \mathbf{E(x) = 0.905}$$
$$\mathrm{I(y,0)} = (2,6) = 0.811$$
$$\mathrm{I(y,1)} = (4,4) = 1 \quad \mathbf{E(y) = 0.905}$$
$$\mathrm{I(z,0)} = (6,2) = 0.811$$
$$\mathrm{I(z,1)} = (0,8) = 0 \quad \mathbf{E(z) = 0.4055}$$

The first splitting variables are done by observing the entropy of all the nodes and selecting the variable which has least entropy as first splitting node. Here, *node z* has least entropy and selected 'z' as the first splitting variable. The Table 2 is divided into two truth tables depending on the conditions when z is 0 and 1.

For $z = 0$ the truth table is:

Table 3. Generation of truth table considering z = 0

w	x	y	F	w	x	y	F
0	0	0	1	1	0	0	1
0	0	1	0	1	0	1	0
0	1	0	0	1	1	0	0
0	1	1	0	1	1	1	0

The truth table when *z* is 0 for all possible combinations of remaining variables is shown in Table 3. The entropy calculation for the remaining variables is shown below. To calculate entropy, refer Entropy calculation of Table 3.

$$\mathrm{I(w,0)} = (3,1) = 0.811$$
$$\mathrm{I(w,1)} = (3,1) = 0.811 \ \mathbf{E(w) = 0.811}$$
$$\mathrm{I(x,0)} = (2,2) = 1$$
$$\mathrm{I(x,1)} = (4,0) = 0 \ \mathbf{E(x) = 0.5}$$
$$\mathrm{I(y,0)} = (2,2) = 1$$
$$\mathrm{I(y,1)} = (4,0) = 0 \ \mathbf{E(y) = 0.5}$$

For $z = 1$ the truth table is:

The Table 4 is generated for all the combinations of remaining nodes when the *node* z is 1. The entropy value for all remaining variables is as shown below.

Table 4. Generation of truth table considering d = 1.

w	x	y	F	w	x	y	F
0	0	0	1	1	0	0	1
0	0	1	1	1	0	1	1
0	1	0	1	1	1	0	1
0	1	1	1	1	1	1	1

To calculate Entropy, refer Entropy calculation of Table 4.

$$\mathrm{I(w, 0)} = (0, 4) = 0$$
$$\mathrm{I(w, 1)} = (0, 4) = 0 \ \mathbf{E(w) = 0}$$
$$\mathrm{I(x, 0)} = (0, 4) = 0$$
$$\mathrm{I(x, 1)} = (0, 4) = 0 \ \mathbf{E(x) = 0}$$
$$\mathrm{I(y, 0)} = (0, 4) = 0$$
$$\mathrm{I\ (y, 1)} = (0, 4) = 0 \ \mathbf{E\ (y) = 0}$$

The next splitting variable is decided which has the least Entropy for two possible combinations of *node* z. Here, the *node* y is selected as next splitting variable.

For $z = 0$; $y = 0$ the truth table is:

Table 5. Generation of truth table considering z = 0 and y = 0.

w	x	F	w	x	F
0	0	1	1	0	1
0	1	0	1	1	0

The next splitting variable is selected, based on the entropy calculations of remaining nodes for all possible combination of two *node* z and y. The Table 5 is generated for the *node* z is 0 and y is 0 for all possible combinations of *node* w and x. The entropy calculations of *node* w and *node* x is shown below.

$$\mathrm{I(w, 0)} = (1, 1) = 1$$
$$\mathrm{I(w, 1)} = (1, 1) = 1 \ \mathbf{E(w) = 1}$$
$$\mathrm{I(x, 0)} = (0, 2) = 0$$
$$\mathrm{I(x, 1)} = (2, 0) = 0 \ \mathbf{E(x) = 0}$$

Table 6. Generation of truth table considering z = 0; y = 1.

W	X	F	W	X	F
0	0	0	1	0	0
0	1	0	1	1	0

For $z = 0$; $y = 1$ the truth table is:

The next splitting variable is selected, based on the entropy calculations of remaining nodes for all possible combination of two *node* z and *node* x. The Table 6 is generated for the *node* z is 0 and y is 1 for possible permutations of *node* w and x. The entropy calculations of *node* w and *node* x is shown below.

$$\begin{aligned}
&\mathrm{I}(\mathrm{w}, 0) = (2, 0) = 0 \\
&\mathrm{I}(\mathrm{w}, 1) = (2, 0) = 0 \quad \mathbf{E(w) = 0} \\
&\mathrm{I}(\mathrm{x}, 0) = (2, 0) = 0 \\
&\mathrm{I}(\mathrm{x}, 1) = (2, 0) = 0 \quad \mathbf{E(x) = 0}
\end{aligned}$$

For $z = 1$; $y = 0$ the truth table is:

Table 7. Generation of truth table considering d = 1; c = 0.

W	X	F	W	X	F
0	0	1	1	0	1
0	1	1	1	1	1

The next splitting variable is selected, based on the entropy calculations of remaining nodes for all possible combination of two *node* z and *node* y. The Table 7 is generated for the *node* $z = 1$ and $y = 0$ for all permutations of w and x. The entropy calculations of w and x is shown below.

$$\begin{aligned}
&\mathrm{I}(\mathrm{w}, 0) = (0, 2) = 0 \\
&\mathrm{I}(\mathrm{w}, 1) = (0, 2) = 0 \quad \mathbf{E(w) = 0} \\
&\mathrm{I}(\mathrm{x}, 0) = (0, 2) = 0 \\
&\mathrm{I}(\mathrm{x}, 1) = (0, 2) = 0 \quad \mathbf{E(x) = 0}
\end{aligned}$$

For $z = 1$; $y = 1$ the truth table is:

The next splitting variable is selected, based on the entropy calculations of remaining nodes for all possible combination of two *node* z and *node* y. The Table 8 is generated for the *node* $z = 1$ and $y = 1$ for all possible permutations of w and x.

Table 8. Generation of truth table considering d = 1; c = 1.

w	x	F	w	x	F
0	0	1	1	0	1
0	1	1	1	1	1

The Entropy calculations of *w* and *x* is shown below.

$$\begin{aligned}
&\mathrm{I(w, 0) = (0, 2) = 0}\\
&\mathrm{I(w, 1) = (0, 2) = 0} \quad \mathbf{E(w) = 0}\\
&\mathrm{I(x, 0) = (0, 2) = 0}\\
&\mathrm{I(x, 1) = (0, 2) = 0} \quad \mathbf{E(x) = 0}
\end{aligned}$$

The 3rd splitting variable is selected based up on the entropy values which have the least value in all possible combinations of z and y. Here, *node x* as the next splitting variable. The Table 8 is generated, based on the Table 2. The decimal equivalents may change, because the corresponding binary equivalents changes, when the variables' order is altered. So, the new truth table (Table 9) with its corresponding decimal equivalents is generated from the main truth table.

In general, representation of any node with value 0/1 is shown by a dotted/solid line. Figure 14 is drawn based on the values in the truth table. This procedure is repeated until the final leaves are displayed as outputs and all of the variables have been completed. Figure 15, When $w = 1$and no matter what the values of nodes *x* and *y* are, *z* represent the same output, which is 1.

Therefore, all of the nodes have been eliminated, and the value of *node z* will be directly associated with the value 1 (See Fig. 16). If the *node w* is LOW and the *node y* is HIGH (see dotted small circle in Fig. 15), then irrespective of the values of node *y*; and *z* corresponds to the same output (LOW), then as a result, node *z* can be removed, and the *node y* can be assigned directly to 0.

When node *x* is 0, all other values of node *z* in the leftmost sub tree/sub graph become 1and similarly, if the value of node *x* is 1, all values of node *z* become 0 Then both the leaves are merged into single output leaf. Figure 17 represents the value of *node w* = 1 when *node x* = 0 and *node w* = 0 when the value of *node x* = 1. Finally, the *node w* is eliminated and the terminal nodes are connected to nod*e x* directly from the OBDD. Figure 17 has three nodes and displays the ROBDD, the final output. Hence, the DSOP from the above BDD are: z + xy'z'.

Binate Covering with Recursion Without DC
The Covering Matrix is:
In the covering matrix, if any variable is not present in disjoint cubes, then that variable is represented with the symbol '2' and is as shown in Table 10 and the corresponding binate select is shown in Fig. 19. The literals can be represented by a mathematical

Table 9. Variable ordered truth table for 4 Variable BF.

z y x w	w x y z	Binary equivalents	F
0 0 0 0	0 0 0 0	0	1
0 0 0 1	1 0 0 0	8	1
0 0 1 0	0 1 0 0	4	1
0 0 1 1	1 1 0 0	12	1
0 1 0 0	0 0 1 0	2	0
0 1 0 1	1 0 1 0	10	0
0 1 1 0	0 1 1 0	6	1
0 1 1 1	1 1 1 0	14	1
1 0 0 0	0 0 0 1	1	0
1 0 0 1	1 0 0 1	9	0
1 0 1 0	0 1 0 1	5	0
1 0 1 1	1 1 0 1	13	0
1 1 0 0	0 0 1 1	3	0
1 1 0 1	1 0 1 1	11	0
1 1 1 0	0 1 1 1	7	1
1 1 1 1	1 1 1 1	15	1

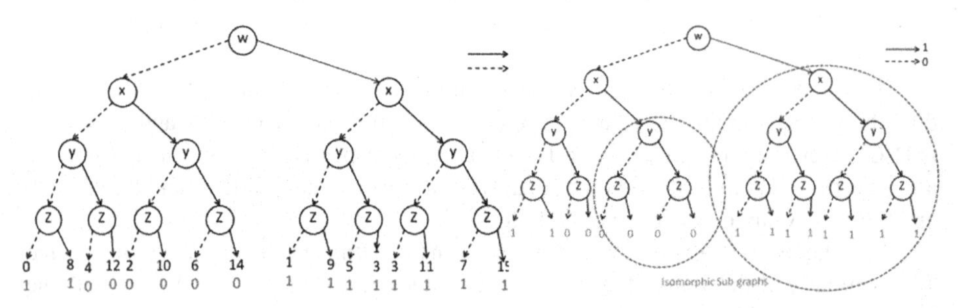

Fig. 14. Generation of variable ordered BDD with the DCs

Fig. 15. Merging the equivalent nodes in variable ordered BDD with the DCs

equation as expressed in (4)

$$u = \begin{cases} 2 \text{ } if \text{ } u \text{ } is \text{ } missing \\ 0 \text{ } if \text{ } u \text{ } is \text{ } a \text{ } complement \\ 1 \text{ } if \text{ } u \text{ } is \text{ } normal \end{cases} \quad (4)$$

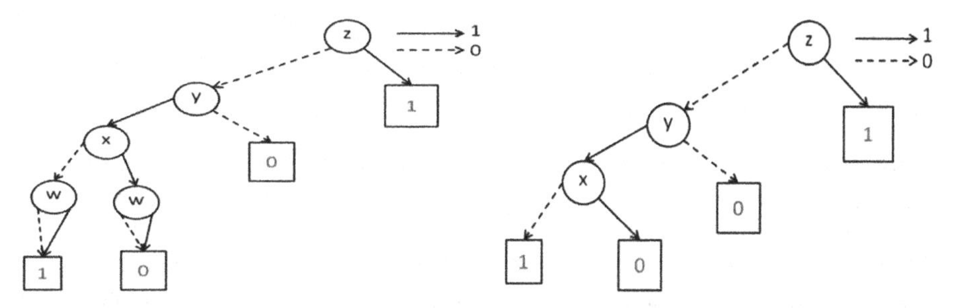

Fig. 16. After merging equivalent invariable OBDD

Fig. 17. ROBDD by considering the DCs

Table 10. Covering matrix for DSOP from variable ordered BDD with DC

				W	x	y	z
	x′	y′		2	0	0	2
x′	y	z		2	0	1	1
x	y	z		2	1	1	1

Binate Select

The binate select is a process in which the first node is the first splitting node (for definitions refer Subsect. 4.4), variable order, and this procedure follows up to the 3 variables because, the example is considered for the 4 variables and cannot be further be minimized. Figure 18 gives an idea of how the binate select is performed with reference to Table 10. The binate select is further continued by the merging process (See Fig. 22) to obtain the reduced SOP.

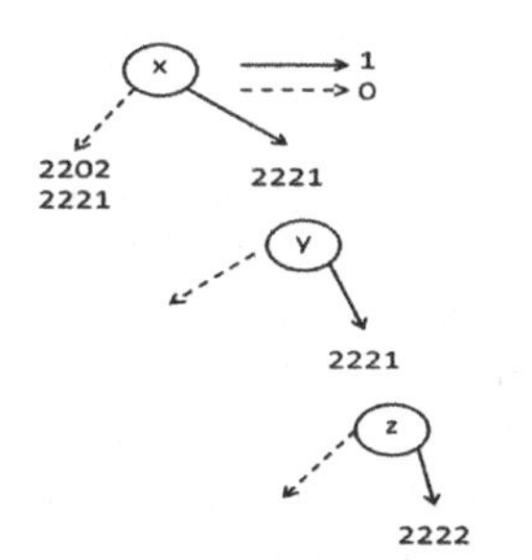

Fig. 18. Binate selecting for DSOP from OBDD without DC

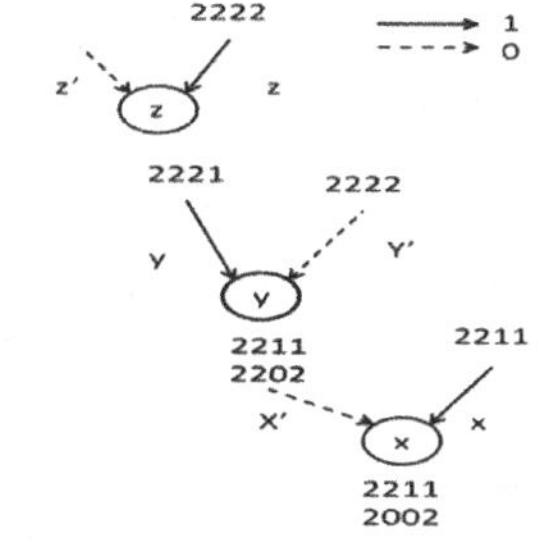

Fig. 19. Merging DSOP from OBDD without DC

Merging

The merging process is done for the nodes which is last node (z) in binate select process. The output of last node (z) in binate select is input to the next node (y/x) of merging. Then, the output changes depending upon the indicating line (dotted line (0)/solid line (1)). In this way, merging is done for all the splitting variables from last node z to the first node x. If the input of node in merged BDD (see Fig. 19) is same as the output of the binate select (see Fig. 19), then the output is same as the input with no change.

Finally, after simulation the obtained simplified expression is $\boldsymbol{F = yz + x'y'}$. To summarize, the given BF with 4 input variables with 8 product terms require 15 nodes for construction of a BDD with 15 logic gates (AND, OR, NOT). There is a possibility that fewer gates will be used when universal gates are used. The given function is further reduced to an expression with two product terms using BDD, requiring three nodes and five gates. From the obtained results it is clear that the cost of SOP implementation reduces, propagation delay decreases as the number of gates reduces. The same procedure is done by considering DC as explained in preceding session.

Binate Covering with Recursion with DC
The Covering Matrix
The Table 11 represents the disjoint cube variables. The literals can be represented by a mathematical equation as expressed in (5).

Table 11. Covering matrix for DSOP from variable Ordered BDD with the DC

					w	x	y	z
			z		2	2	2	1
	x′	y′	z′		2	0	0	0

Binate Select
The binate select is a process in which the first node is the first splitting node that determines the order of the variables. This procedure continues until there are 3 variables because the example takes into account the DCs for 4 variables.

Merging
The merging process is first done for the node which is the last node (x) in binate select process. The output of the last node x in binate select is the input to the next node (y/z) of merging. Then, the output changes depending on the indicating line (dotted line (0)/solid line (1)). In this way, merging is done for all the splitting variables from last node x to the first node z. If the input of node in merged BDD (see Fig. 21) is same as the output of the binate select (see Fig. 20), and then the output is same as the input with no change.

Finally, after simulation the obtained simplified expression is $\mathrm{F} = \mathrm{z} + \mathrm{x}'\mathrm{y}'$. To summarize, the given Boolean function with 4 input variables with 8 product terms and two DC product terms require 15 nodes for construction of a BDD with 17 logic gates (AND, OR, NOT). Using of BDD, the given function reduces to an expression with two product terms and this requires 3 nodes with 4 gates. From the obtained results it is clear that the cost of SOP implementation reduces, propagation delay decreases as the number of gates reduces with the use of BDD than the conventional approaches.

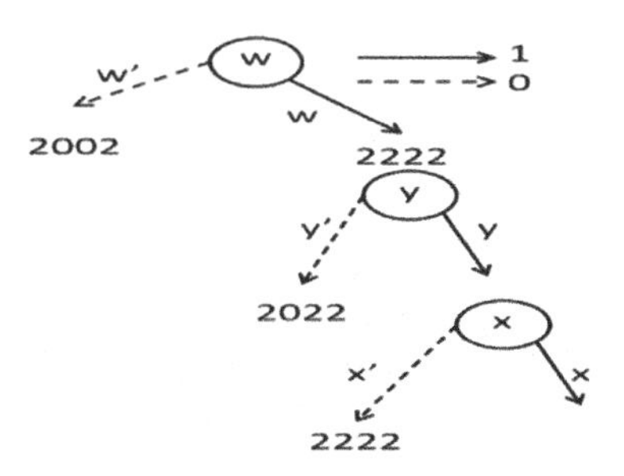

Fig. 20. Binate select for DSOP from BDD with DC

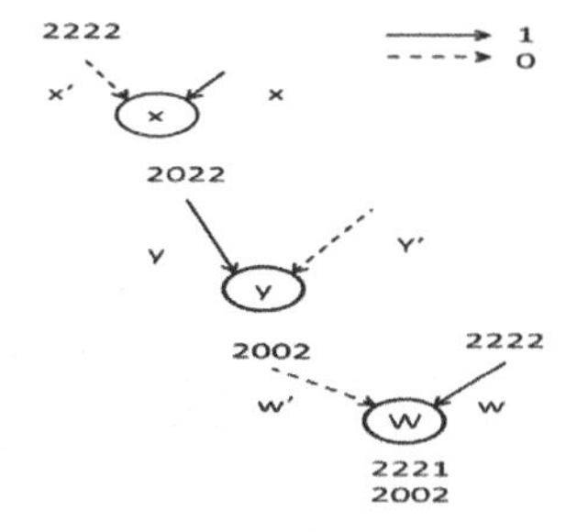

Fig. 21. Merging DSOP from BDD with DC

6 Simulation Results

The simulation results are the results obtained from the written code for the examples considered with and without DCs in session 2. The process of applying the steps to Binary decision diagram to convert to the reduced binary decision diagram is provided as follows.

6.1 Boolean Expression with the DC

The BF with DC (5) is considered as an example

$$F = (0, 1, 3, 7, 8, 9, 11, 15) + d(5, 13) \tag{5}$$

$$\begin{aligned} F =& w'x'y'z' + w'x'y'z + w'x'yz + w'xyz + wx'y'z' + wx'y'z \\ &+ wx'yz + wxyz + w'xy'z + wxy'z \end{aligned} \tag{6}$$

The Boolean function is expressed as Boolean expression (5). The Fig. 22 shows the BDD of the given SOP created for the all possible combinations of input variables as 0 and 1. The output of the sink node indicates the function outputs.

The reduction rules reduce the redundant nodes and merge equivalent nodes and are applied to the BDD (Fig. 22), and then the resultant output (Fig. 23) is the ROBDD. The variable OBDD is obtained in the same manner as discussed in afore mention sections to get variable ordered BDD (Fig. 24); reduced variable ordered BDD (Fig. 25).

6.2 Expressions Considered for 3 and 4 Variables (with & Without DC)

The expressions considered for 3, 4 variables with and without DC has been taken to generate the Tables 12, 13, 14 and 15 for number of nodes and logic gates for all possible variable orders.

Expression for 3 Variables Without DC. Boolean Function: $F = \sum m(0, 1, 2, 4, 6)$; Boolean Expression: $F = x'y'z' + x'y'z + x'yz' + xy'z' + xyz'$

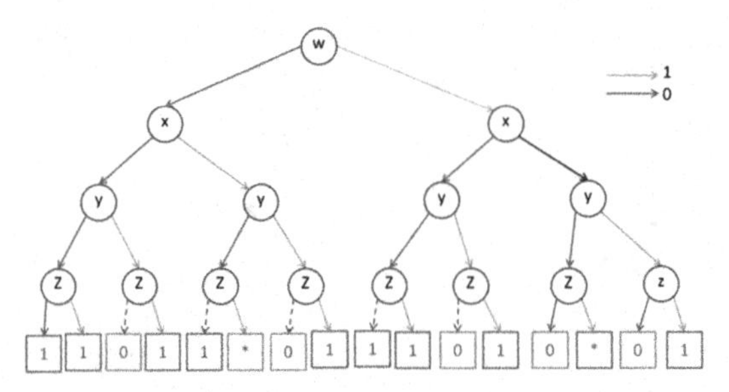

Fig. 22. BDD for SOP with DC.

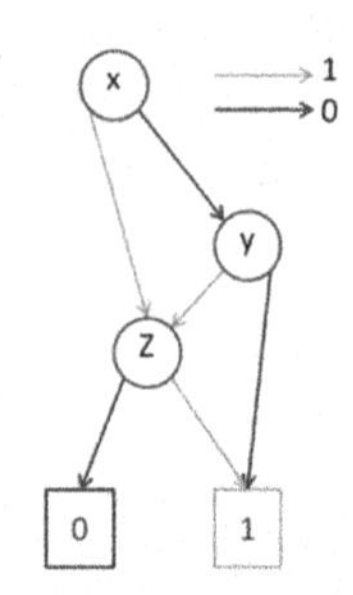

Fig. 23. ROBDD with DC

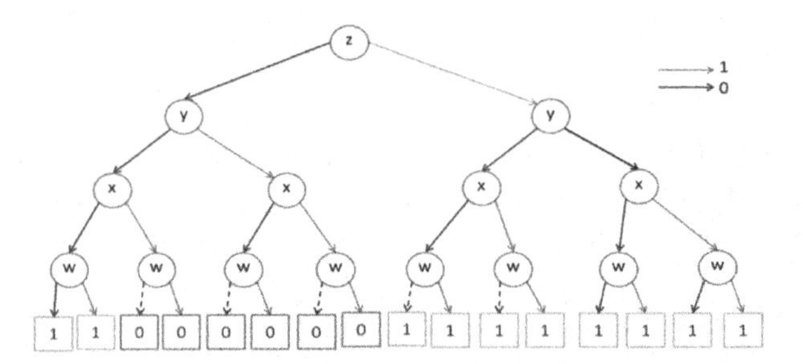

Fig. 24. Variable OBDD with DC

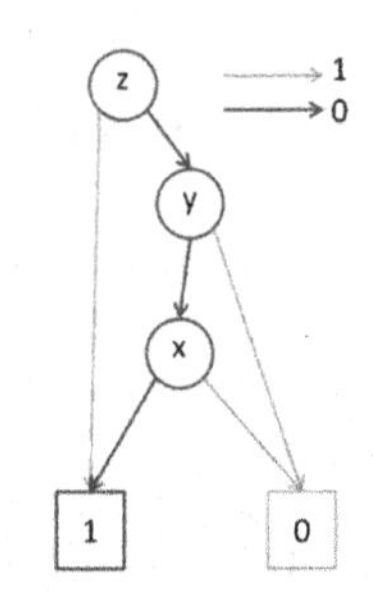

Fig. 25. ROBDD of Fig. 24 with DC

Table 12. Results for 3 variable Boolean expression without dc

$F = x'y'z' + x'y'z + x'yz' + xy'z' + xyz'$		
Variable order	Minimization using BDD	
	Number of nodes in ROBDD	Number of logic gates
xyz	5	7
xzy	5	7
yzx	5	7
yxz	5	7
zxy	5	7
zyx	5	7

Without simplifying, the number of nodes and logic gates required for the given Boolean function is as follows:

$$\text{Number of Gates} = 11; \text{Number of Nodes} = 7$$

Table 12 shows that the variable order used in the BDD approach that determines the number of logic gates and nodes in ROBDD. For all possible combinations (No. of combinations 3! = 6) of 3 variables SOP without the DC the order of the number of nodes and logic gates in ROBDD is as mentioned in Table 12. Table 12 exhibits that the variable order has no effect on the number of logic gates and nodes. Hence, any variable order may be considered for this particular case.

Expression for 3 Variables with DC. Boolean Function: $F = \sum m(0, 1, 2, 4, 6) + d(5)$; Boolean Expression $F = x'y'z' + x'y'z + x'yz' + xy'z' + xy'z + xyz'$.

Without simplifying, the number of nodes and logic gates required for the given Boolean function is as follows:

$$\text{Number of gates} = 12; \quad \text{Number of nodes} = 7$$

According to Table 13, the number of logic gates and nodes in ROBDD are proportional to the variable order when employing the BDD approach. For all possible combinations (No. of combinations $3! = 6$) of 3 variables SOP with the DC, for different variable order, the number of nodes and logic gates in ROBDD are same as referenced in Table 13. According to Table 13, the number of logic gates and nodes remains constant regardless of the variable order. Moreover, it is understood that the variable order (*xyz*; *xzy*; *yzx*; *yxz*) the logic circuit may need 4 gates with 4 nodes in the ROBDD. Hence, any variable order may be considered for this particular case and would be the best order. The variable order (*zxy*; *zyx*) are coined as *worst order* because more nodes are present after reduction of BDD though the number of logic gates are same.

Table 13. Results for 3 variable Boolean expression by considering DCs

$F = x'y'z' + x'y'z + x'yz' + xy'z' + xyz'$		
Variable order	Minimization using BDD	
	Number of nodes in ROBDD	Number of logic gates
xyz	4	4
xzy	4	4
yzx	4	4
yxz	4	4
zxy	4	4
zyx	5	4

Expression 4-variable without DC. Boolean Function: $F = \sum m(7, 10, 12, 14, 15)$.

Boolean expression $F = w'xyz + wx'yz' + wxy'z' + wxyz' + wxyz$.

Without simplifying, the number of nodes and logic gates required for the given Boolean function is as follows:

$$\text{Number of gates} = 11; \quad \text{Number of nodes} = 15$$

For every possible combination (No. of combinations $4! = 24$) of the variables, the number of nodes and logic gates in ROBDD required are mentioned in the Table 14. The best variable order obtained using entropy calculation is '*wxyz*' because it requires less number of nodes (6) and logic gates (4) compared to other combinations in ROBDD. The worst variable order is '*wxyz*' and '*wyzx*' because it requires more number of logic gates (8) and nodes (8) for expression without DC. While remaining all other combinations mostly the number of gates (6) with 7 ROBDD nodes.

Table 14. Results for 4 variable Boolean expression without DCs

$F = w'xyz + wx'yz' + wxy'z' + wxyz' + wxyz$					
Variable order	Minimization using BDD		Variable order	Minimization using BDD	
	No. of nodes in ROBDD	No. of Logic Gates		No. of nodes in ROBDD	No. of Logic Gates
w x y z	8	8	y w z x	7	6
w x z y	6	4	y w x z	7	6
w yx z	7	5	y x w z	8	7
w y z x	8	8	y x z w	7	6
w zx y	7	7	y z x w	7	6
w z y x	8	4	y z w x	7	6
x y z w	8	7	z w x y	7	6
x y w z	7	6	z w y x	7	6
x wy z	7	6	z x w y	7	6
x w z y	7	6	z x y w	7	6
x z w y	7	6	z y w x	7	5
x z y w	7	6	z y x w	7	7

Expression 4-variable with DC. Boolean Function:

$$F = \sum m(7, 10, 12, 14, 15) + d(13).$$

Boolean expression $F = w'xyz + wx'yz' + wxy'z' + wxyz' + wxyz + wxy'z$.

The number of nodes and logic gates required for the given Boolean function without simplification is as given below:

Number of gates = 12; Number of nodes = 15

For every possible combination (No. of combinations $4! = 24$), the number of nodes and logic gates in ROBDD required are obtained as mentioned in the Table 15 with DC. The best variable order obtained using entropy calculation is 'wxyz' and 'wyxz' because it requires less number of nodes (6) and logic gates (4) compared to other combinations in ROBDD. The worst variable order are at most 9 combinations (xyzw, xywz; xzwy; xzyw; yxzw; yzxw; yzwx; zxyw and zywx) because it requires more number of logic gates (7) and nodes (8) for expression without DC.

Table 15. Results for 4 variable Boolean expression with DCs

$F = w'xyz + wx'yz' + wxy'z' + wxyz' + wxyz$

Variable order	Minimization using BDD		Variable order	Minimization using BDD	
	No. of nodes in ROBDD	No. of logic gates		No. of nodes in ROBDD	No. of logic gates
w x y z	7	6	y w z x	7	6
w x z y	**6**	**4**	y wx z	7	6
w yx z	**6**	**4**	y x w z	7	6
w y z x	7	5	y x z w	8	7
w zx y	7	5	y z x w	8	7
w z y x	7	6	y zw x	8	7
x y z w	8	7	z w x y	7	6
x y w z	8	7	z w y x	7	6
x wy z	7	6	z x w y	8	6
x wz y	7	6	z x y w	8	7
x z w y	8	7	z y w x	8	7
x z y w	8	7	z y x w	6	5

6.3 Graphical Representation of Results

The program was implemented in Intel Corei3, Inspiron 14, 6006U CPU, 2 GHz processor with 4 GB RAM with Python 2.0.1Student Edition platform. The simulation results are explained in detail. As the number of input variables increased, the amount of time required in generating the BDD increased as well (the graph was not provided) and Figs. 26 and 27 depict the reduction in number of logic gates and nodes using BDD for 3 and 4 Variable BF.

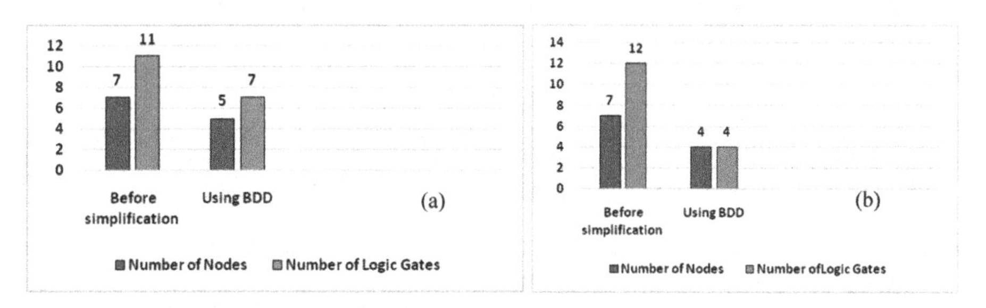

Fig. 26. Comparison of No. of Nodes and No. of Logic gates for a 3 variable Boolean function before simplification and after using BDD (a) without DC (b) with DC

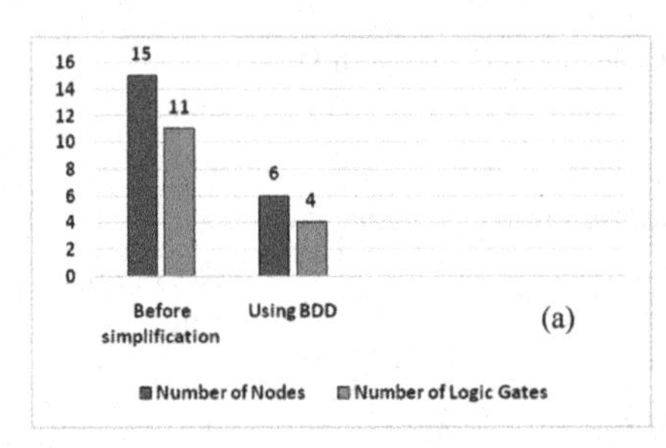

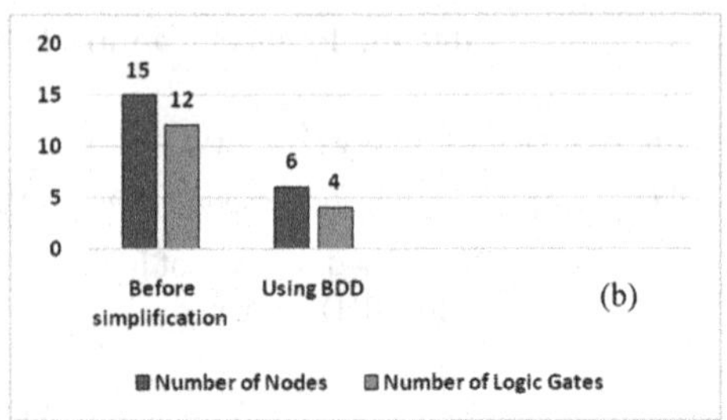

Fig. 27. Comparison of No. of Nodes and No. of Logic gates for a 4 variable Boolean function before simplification and after using BDD (a) without DC (b) with DC

7 Conclusion

The paper provides a detailed understanding of the minimization of the Boolean functions (a) without DC (b) with DC using BDD. The results indicate that BDD is a good alternative for SOP Boolean function reduction as it takes less number of nodes for the logic circuit implementation; less propagation delay, low power consumption. In this work, an algorithm has been developed to minimize the Boolean expression using BDD with the aid of reduction and merging rules. The number of nodes is counted, each node's entropy is calculated, and the variable is then arranged in ascending entropy order. Using, covering matrix, binate select and merging techniques, the ROBDD is obtained based on the best variable order. The observations of the obtained results are summarized as follows- The more input variables there are in the BF, the longer it takes to generate the BDD. The number of nodes and logic gates are almost same for the lower inputs (say 3 variable BF) for any variable order, while 50% reduction in number of gates are got when 4- input variables is considered. To learn more about the propagation delay, power consumption, this approach may be validated using a suitable VLSI tool to obtain optimized results.

References

1. Yang, C., Ciesielski, M.: BDS-a BDD-Based Logic Optimization System, TR-CSE-00-1, pp. 1–25 (2000)
2. Minna, L., Jo, S-Y.: A minimization technique for BDD based on micro canonical optimization. The Kips Trans.: Part A **8**(1), 48–55 (2001)
3. Shannon, C.E.: A mathematical theory of communication. The Bell Syst. Tech. Jour. **27**(3), 379–423 (1948)
4. Shannon, C.E., Weaver, W.W.: The Mathematical Theory of Communication. University of Illinois Press (1949)
5. Sensarma, D., Subhashis, B., Krishnendu, B., Saptarshi, N., Samar, S.: On an optimization technique using binary decision diagram. Int. J. Comput. Sci., Eng. Appl.. **2**(1), 73–86 (2012)
6. Banerjee, S., Sensarma, D., Basuli, K., Naskar, S., Sarma, S.S.: The reconstruction conjecture. The Second Int. Conf. Comput. Sci. Inform. Technol. **86**(3), 17–25 (2011)
7. Popel, D.: Towards Efficient Calculation of Information Measures for Reordering of Binary Decision Diagrams. In: Proceedings IEEE International Symposium on Signals Circuits and Systems, pp. 509–512 (2001)

8. Swamy, G.M.: An exact logic minimizer using implicit binary decision diagram based methods. In: ICCAD'94 Proceedings of the 1994 IEEE/ACM international conference on Computer aided design (1994)
9. Fey, G., Drechsler, R.: Utilizing BDDs for disjoint SOP minimization. In: the 2002 45th Midwest Symposium on Circuits and Systems, vol. 2, pp. 306–309 (2002)
10. Bryant, R.E.: Symbolic boolean manipulation with ordered binary-decision diagrams. ACM Comput. Surv. **24**(3), 293–318 (1992)
11. Drechsler, R., Sieling, D.: Binary decision diagrams in theory and practice. Software Tools Technol. Transfer. **3**, 112–136 (2001)
12. Falkowski, B.J.: Spectral Methods for Boolean and Multi-valued Input Logic Functions, Dissertation. Portland State University (1991)
13. Falkowski, B.J., Schafer, I., Chang, C.-H.: An effective computer algorithm for the calculation of disjoint cube representation of boolean functions. In: Proceedings of 36th Midwest Symposium on Circuits and Systems, vol. 2, pp. 1308–1311 (1993)
14. Mishchenko, A., Perkowski, M.: Fast Heuristic Minimization of Exclusive-Sum-of-Products. In: International workshop on Applications of the Reed-Muller Expansion in Circuit Design, pp. 242–250 (2001)
15. Astola, J., Stankovic, R.: Fundamentals of Switching Theory and Logic Design – A Hands on Approach. Springer (2006)
16. Mano, M., Ciletti, M.D.: Digital Logic And Computer Design, 4th edn. Pearson PHI (2016)
17. Kohavi, Z., Jha, N.K.: Switching And Finite Automata Theory, 3rd edn. Cambridge University Press (2009)
18. Godse, A.P., Godse, D.A.: Switching Theory and Logic Design. Technical publications Pune (2009)
19. Kumar, A.A.: Fundamentals of Digital Circuits, 4th edn. PHI Learning Publisher (2016)
20. Lee, C.Y.: Representation of switching circuits by binary decision programs. Bell Syst. Technol. **38**(4), 985–999 (1959)
21. Akers, S.B.: Binary decision diagrams. IEEE Trans. Comput. **C–27**(6), 509–516 (1978)
22. Bryant, R.E.: Graph-based algorithms for boolean function manipulation. IEEE Trans. Comput. **35**(8), 677–691 (1986)
23. Bryant, R.E.: On the complexity of VLSI implementations and graph representations of Boolean functions with application to integer multiplication. IEEE Trans. Comput. **40**(2), 205–213 (1991)
24. Coudert, O., Madre, J.C., Berthet, C.: Verifying temporal properties of sequential machines without building their state diagrams. In: Clarke, E.M., Kurshan, R.P. (eds.) CAV 1990. LNCS, vol. 531, pp. 23–32. Springer, Heidelberg (1991). https://doi.org/10.1007/BFb0023716
25. Clarke, E.M., Kurshan, R.P. (eds.), American Mathematical Society, pp. 75–84 (1990)
26. Bryant R.E.: Symbolic Boolean Manipulation with Ordered Binary Decision Diagrams. School of Computer Science, Carnegie Mellon University, Pittsburgh, PA, vol. 24(3), 273–318 (1992)

Image Processing

Single Image Dehazing Through Feed Forward Artificial Neural Network

K. Soni Sharmila[1], A. V. S. Asha[1], P. Archana[1], and K. Ramesh Chandra[2(✉)]

[1] Department of CSE, Shri Vishnu Engineering College for Women, Bhimavaram, India
[2] Department of ECE, Vishnu Institute of Technology, Bhimavaram, India
rameshchandra001@gmail.com

Abstract. Due to light scattering by air particles, images taken under bad lighting conditions (such as haze, fog, mist, or smog) have a reduced level of visibility. One blurry image can be made visible again using single image dehazing techniques. Due to the ill-posedness of the Single image dehazing problem, it is difficult to solve. Even though, the Dark channel prior (DCP) has been the most prevalent method for image dehazing algorithms, but it suffers from huge computational time and picture quality. The accurate transmission map estimation is one of the best preferred ways to achieve the least computational time and high picture quality in DCP. Hence, this work is focused on constructing transmission map of a hazy image based on feed forward artificial neural network. The recommended method uses a feed forward ANN to conduct the transmission map straight from the minimum channel and a normalizing procedure to increase the re-covered image information. With a training set of eighty images, the network is trained by means of mean square error (MSE). The proposed method utilizing peak signal to noise ratio (PSNR) and structural similarity (SSIM) index measures are used to assess the restoration quality. The investigational conclusions have demonstrated that, the suggested method outperforms the dehazing of an input image without degrading visible quality (PSNR = 69.16, SSIM index = 0.8913). In addition, the proposed method is suited for real-time applications given the average computing time it achieves is 1.03 s.

Keywords: Dark channel prior · Dehazing · Transmission map · Multilayer perceptron · Restoration

1 Introduction

Haze is a normal environmental oddity that is caused by minute particles like dirt, fume, and airborne fog, which reduces the clarity of the picture. Dehazing calculations were traditionally used for a narrow range of purposes since experts considered it as a photo handling technique to recover image details. But the requirements of significant level computer vision assignments, the rapid advancement of independent frameworks, and artificial insight have all led to a renewed interest in more advanced picture dehazing

N. Gupta et al. (Eds.): IC4S 2022, LNICST 472, pp. 115–124, 2023.
https://doi.org/10.1007/978-3-031-28975-0_9

techniques [1]. Dehazing estimation improvement has thus become possibly the most important topic of artificial intelligence, and one of its most fundamental applications is the use of dehazing computations to advance the display of autonomous structures and stages for adverse barometer conditions.

Most often, a computer vision framework's optics are designed under the assumption that there will be beautiful weather patterns, in which case each pixel color intensity will only be related to the brightness of the initial scene. As a result, learners in the early stages of computer vision tasks specifically missed the state of bad weather [2]. However, experts rapidly realized the value of photo reclamation techniques. Outside images are unavoidably adversely impacted by the environment, and refraction, scattering, and maintenance do occur even on reasonably sunny mornings resulting in the loss of specific information and low differentiation. Unfavorable consequences on independent frameworks are undoubtedly caused by these degraded images [3]. Figure 1, depicts the basic atmospheric scattering model in which the haze is formed on the images.

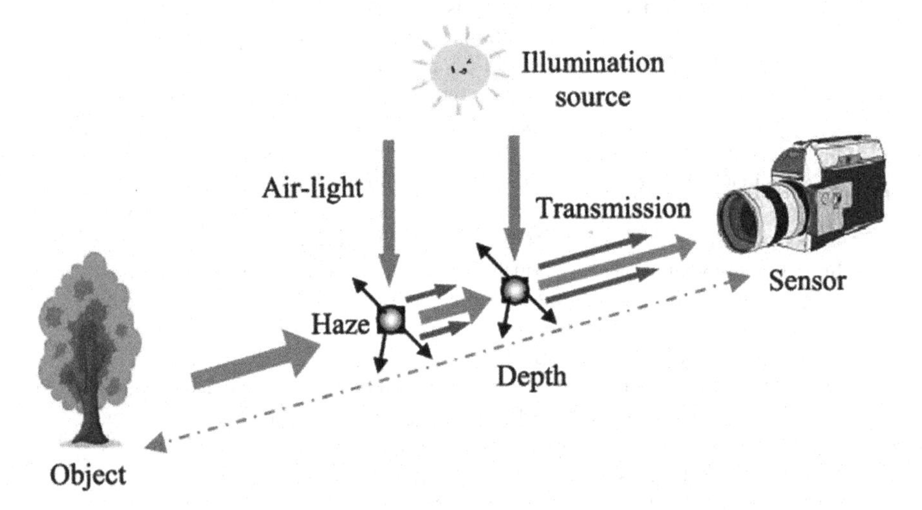

Fig. 1. Haze creation model.

Haze ejection is particularly appealing in a variety of industries, including computer vision analyses, image processing, and visual art [4, 5]. To begin with, the visibility of a dim image brought on by air particles is improved when the haze is removed from the image. The scene is the related image. Luminant is used in the majority of computerized image processing techniques, including sophisticated scale shape identification. The scenes are what are used to evaluate these techniques. When the image or picture is tedious, visual computations struggle and struggle to perform well. In this way, removing the fog is necessary for improved results and productivity. The likelihood of using the horrible images increases [6].

The term spatial variation refers to the corruption that results from scattering that is dependent on the distance of the camera from the scene. The capacity to perceive the dim picture caused by the air particles is increased by the removal of murkiness from the image [7]. Therefore, the focus of this work is on creating an algorithm for image dehazing that has high accuracy and requires minimal calculation time while maintaining acceptable picture quality [8]. This can be achieved by employing the perfect transmission map estimation in DCP algorithm, which can be accomplished through feed forward ANN. Here, the transmission map has been approximated through a minimum channel without lowering the image quality and the network has been trained using a data set of 80 images with MSE.

2 Related Work

Most single-image dehazing computations use learning-based highlights or previously computed picture handling highlights to predict transmission, whereas exact techniques are used to measure the barometric light (the exemption being start to finish learning-based calculations). However, recently the single image dehazing based on convolutional neural network (CNN), artificial neural networks (ANN), and deep neural network (DNN) has attained enormous significance among the research community as they have offered less computational time and high accuracy when compared to conventional dehazing algorithms (DCP, Fast DCP).

Using clear photos as positive instances and hazy images as negative instances, a unique contrastive regularization (CR) based on contrastive learning has been proposed [9] which is used to take advantage of both the information contained in each type of image. CNN grounded Deep architecture has been presented [10], and its layers are specifically created to encapsulate the established presumptions/priors in picture dehazing. The authors in [11] also developed a CNN based dehazing algorithm. To unswervingly rebuild the dehaze image, the authors of [12] presented an endwise feature fusion attention network (FFA-Net).

Image haze removal techniques traditionally rely on estimating a transmission map. The lack of depth information makes this a poorly presented challenge when dealing with single photos. The authors in [13] proposed an end-to-end learning-based method for directly removing haze from an image by employing a modified conditional generative adversarial network. A GAN for a single image that is cycle-consistent dehazing method called CD-Net has been suggested; it is trained on a dataset of real-world hazy images in an unpaired fashion [14].

For single-image dehazing, the authors of [15] developed a particular generative adversarial network (GAN). The authors in [16] were driven to consider the Laplacians of Gaussian (LoG) of the images, which astonishingly preserves this information, in order to address the problem of single image haze removal due to the challenges in single picture dehazing. According to the authors of [17], ignoring a continual atmospheric light guess, a unique dehazing network that concurrently estimates the transmission map has been proposed.

A unique method to eliminate haze degradations in RGB photos has been developed by the authors in [18] utilizing a layered conditional Generative Adversarial Network (GAN). To remove the haze on each color channel separately, it uses a triplet of GAN. Numerous authors have developed single image dehazing algorithms based on CNN, DNN and GAN, which highlights the effective estimation of the transmission map of dark channel image. Numerous new investigations based on Arduino integrated with

IOT [19–23] perceive the requirement for ongoing handling and the advancement of methods to lessen the capacity, intricacy, handling time and other related angles without compromising the presentation. Hence, in this work a feed forward artificial neural network has been proposed for effectively estimating the transmission map of hazy image without degrading image quality by a means of trained data set of eighty images.

3 Methodology

Usually, the DCP consists of four major steps for image dehazing such as; atmospheric light estimation; transmission map estimation; transmission map refinement; reconstruction. The major disadvantage of DCP methodology is its computational time. Introducing ANN in transmission map refinement stage has reduce the computational time.

3.1 Atmospheric Light Estimation

The haze image is a combination of image coordinates, observed image, haze free image, atmospheric light, scattering coefficient and scene depth [3], which is represented using (1). The dark channel from the input hazy image is estimated by the majority of traditional DCP-based dehazing techniques and it is represented using (2).

$$I_{haze}(x) = J_{haze-free}(x)e^{-\beta d(x)} + A\left(1 - e^{-\beta d(x)}\right) \tag{1}$$

$$J^{dark}(x) = \min_{y \in \Omega(x)} \left(\min_{c \in \{r, g, b\}} J^c(y) \right) \tag{2}$$

3.2 Transmission Map Estimation

From Eq. (1), β is the atmospheric scattering coefficient and d is the scene depth. The transmission map has been estimated [3] with β, and d values and it is represented using eq. (3).

$$t(x) = e^{-\beta d(x)} \tag{3}$$

3.3 Transmission Map Refinement

False textures and blocking artefacts might result from improper estimate of the transmission map. Hence, transmission map refinement based on feed forward artificial neural network has been considered for accurate refinement. The multilayer perceptron approach, which was developed from neurons, is utilized in transmission map refining. The basic building blocks of the brain and nervous system are biological neurons, often known as nerve cells or just neurons. These are the cells that receive sensory input from the outside world via dendrites, analyses it, and then transmit it to other cells via axons. The feed forward neural network is supplemented by the multi-layer perceptron (MLP).

A data set of 80 images having mean square values have been trained in perceptron which are having the dimensions of 512 × 512 and having different picture quality. The transmission map has been evaluated flowed by preprocessed dark channel construction. As depicted in Fig. 2, it has three different sorts of layers: input layer, output layer, and hidden layer [24].

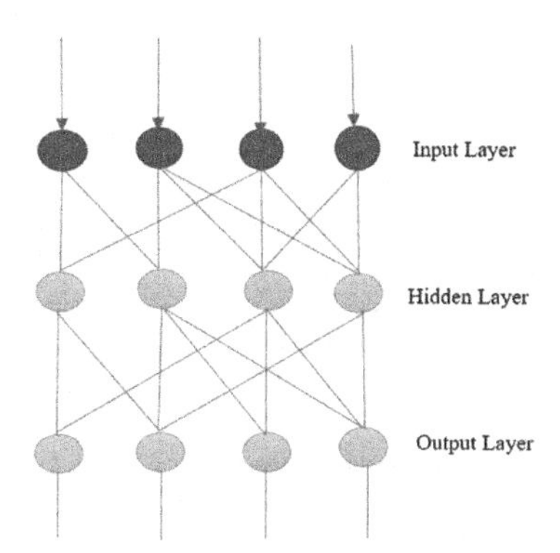

Fig. 2. Representation of Feed Forward ANN with Single Hidden Layer [24].

The signal is received at the input layer, where it will be processed. The output layer completes the necessary tasks, such as classification and prediction. The true computational engine of the MLP is composed of an arbitrary number of hidden layers sandwiching the input and output layers. Similar to a feed forward network, data moves forward from an MLP's input layer to its output layer. The back propagation learning approach is used to train the MLP's neurons [24]. A perceptron receives n input features ($x = x_1, x_2, \ldots, x_n$), and each one is given a weight (see Fig. 3). It is vital to have features for numerical input. As a result, in order to use a perceptron, nonnumeric input properties must be transformed into numeric ones.

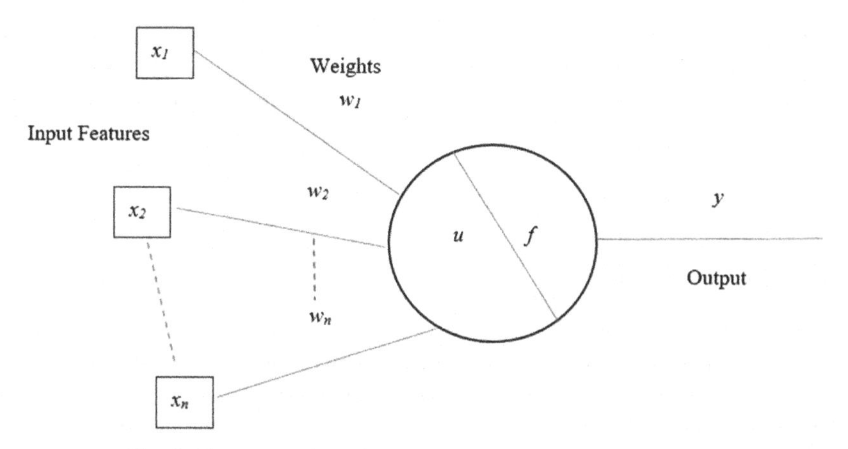

Fig. 3. Representation of Perceptron with *n* input features [24].

During pathway, the starting layer has been biased and applied with data stream, and it is continued for all layers until accomplished output has been formed (see Fig. 4). After comparing the network's actual and predicted output values, a fault signal is intended for each output node [25].

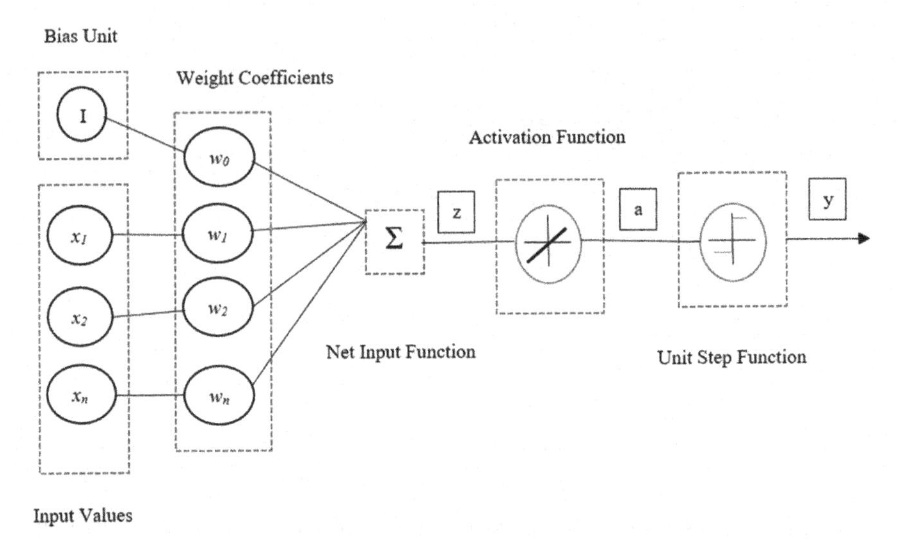

Fig. 4. Feed Forward ANN Architecture with Single Layer [26].

For all hidden layers, until each weight in the system attained a fault signal representing its comparative fault. Then, for each weight the faults are updated until the training patterns are fully encoded.

4 Simulation Results

First, the proposed approach has been applied to the input haze image. It is first applied for dark channel building after being down sampled by four. The dark channel formation was involved in a number of DCP-related processes, and the output was zero-padded. Transmission map has been created following dark channel construction (which is pre-processed). Using a feed forward ANN in the transmission map refining stage has solved the issues with transmission map estimation.

4.1 Subjective Analysis

Subjective analysis is the analysis which uses the observers to make the quality estimation based on their visual opinion of the image. The subjective analysis has been carried out for various hazy images. The simulation results clearly indicates that, the photographic eminence (see Fig. 5) of dehazed images is far better than the input haze image.

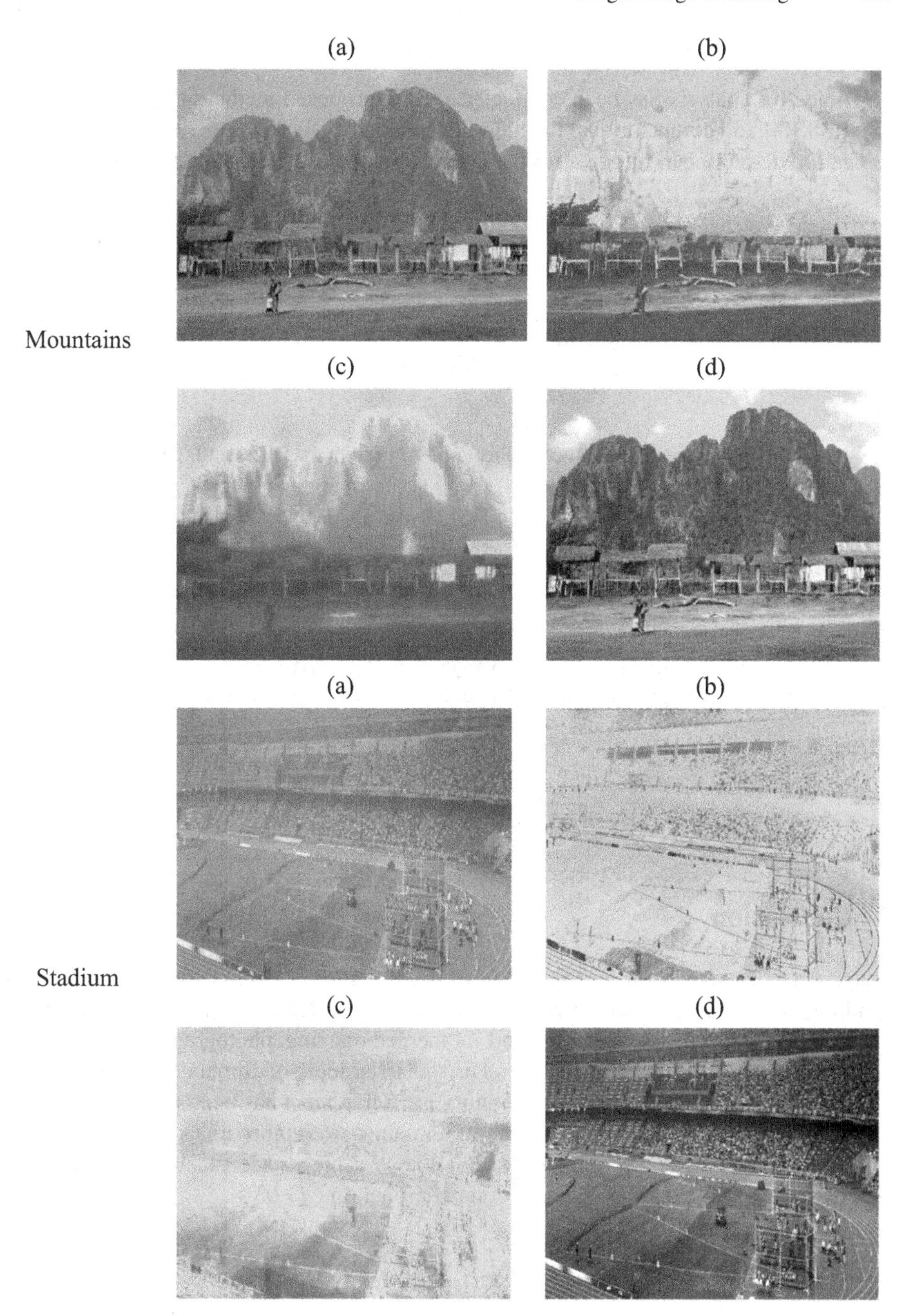

Fig. 5. Simulation Results of Proposed Dehazing Method for different input hazy images (a) Input Hazy Image (b) Initial Transmission Map (c) Feed Forward ANN based Transmission Map (d) Dehazed Image.

4.2 Objective Analysis

The objective analysis has been conducted for the proposed methodology in terms of PSNR, SSIM and computational time. The detailed analysis has been depicted in Table 1. The PSNR and mean square error has been evaluated using (4) and (5) respectively.

$$PSNR = 20\log(\frac{Max}{\sqrt{MSE}}) \tag{4}$$

$$MSE = \frac{1}{MM}\sum_{i=0}^{m-1}\sum_{j=0}^{n-1}[I(i,j) - K(i,j)]^2 \tag{5}$$

Table 1. Performance Measures of Proposed Algorithm for Different Input Images.

Image name	PSNR	SSIM	Computational time (s)
Mountains	68.41	0.9105	1.15
Pumpkins	62.10	0.7469	1.25
Stadium	68.69	0.8695	1.18
Park	61.61	0.7922	1.03
City	62.85	0.6727	1.20
Florence	69.16	0.8913	1.73
Cones	67.66	0.8213	1.08

5 Conclusion

Image dehazing expulsion techniques have become more treasured for many picture handling and computer vision implementations. All the dehazing methods helpful for reconnaissance, for distant detecting and submerged imaging, photography and so forth A large portion of the techniques depend on the assessment of climatic light and transmission map. In, this work the transmission map refinement has been evaluated based on feed forward ANN which is having a train data set of eighty images with normalized mean square error. The proposed methodology achieves acceptable image quality (PSNR-69.16; SSIM-0.9105) and less computational time (1.03 s).

References

1. Kermani, E., Asemani, D.: A robust adaptive algorithm of moving object detection for video surveillance. EURASIP J. Image Video Process. **2014**(1), 1–9 (2014)
2. Ozaki, M., Kakimuma, K., Hashimoto, M., Takahashi, K.: Laser-based pedestrian tracking in outdoor environments by multiple mobile robots. In: Proceedings of Annual Conference on IEEE Industrial Electronics Society, pp. 197–202. IECON, Melbourne (2011)

3. Lee, S., Yun, S., Nam, J.-H., Won, C.S., Jung, S.-W.: A review on dark channel prior based image dehazing algorithms. EURASIP J. Image Video Process. **2016**(1), 1–23 (2016)
4. Anwar, M.I., Khosla, A.: Vision enhancement through single image fog removal. Eng. Sci. Technol., An Int. J. **20**(3), 1075–1083 (2017)
5. Badhe, M.V., Prabhakar, L.R.: A survey on haze removal using image visibility restoration technique. Int. J. Comput. Sci. Mob. Comput. **5**(2), 96–101 (2016)
6. Cai, B., et al.: Dehazenet: An end-to-end system for single image haze removal. IEEE Trans. Image Process. **25**(11), 5187–5198 (2016)
7. Fattal, R.: Single image dehazing. ACM Trans. Graphics **27**(3), 1–9 (2008)
8. Gibson, K.B., Vo, D.T., Nguyen, T.Q.: An investigation of dehazing effects on image and video coding. IEEE Trans. Image Process. **21**(2), 662–673 (2012)
9. Wu, H., et al.: Contrastive learning for compact single image dehazing. In: 2021 IEEE/CVF Conference on Computer Vision and Pattern Recognition (CVPR), pp. 10546–10555 (2021)
10. Cai, B., Xu, X., Jia, K., Qing, C., Tao, D.: DehazeNet: an end-to-end system for single image haze removal. IEEE Trans. Image Process. **25**(11), 5187–5198 (2016)
11. Liu, Z., Xiao, B., Alrabeiah, M., Wang, K., Chen, J.: Single image dehazing with a generic model-agnostic convolutional neural network. IEEE Signal Process. Lett. **26**(6), 833–837 (2019)
12. Qin, X., Wang, Z., Bai, Y., Xie, X., Jia, H.: FFA-Net: Feature fusion attention network for single image dehazing. Proc. AAAI Conf. Artif. Intell. **34**(07), 11908–11915 (2020)
13. Bharath Raj, N., Venketeswaran, N.: Single image haze removal using a generative adversarial network. In: 2020 International Conference on Wireless Communications Signal Processing and Networking (WiSPNET), pp. 37–42 (2020)
14. Dudhane, A., Murala, S.:CDNet: single image de-hazing using unpaired adversarial training. In: 2019 IEEE Winter Conference on Applications of Computer Vision (WACV), pp. 1147-1155 (2019)
15. Zhu, H., et al.: Single-image dehazing via compositional adversarial network. IEEE Trans. Cybern. **51**(2), 829–838 (2021)
16. Kumar, P., Priyankar, S., Sur, J.A.:Scale-aware conditional generative adversarial network for image dehazing. In: 2020 IEEE Winter Conference on Applications of Computer Vision (WACV), pp. 2344–2354 (2020)
17. Zhang, H., Sindagi, V., Patel, V.M.: Joint transmission map estimation and dehazing using deep networks. IEEE Trans. Circuits Syst. Video Technol. **30**(7), 1975–1986 (2020)
18. Suárez, P.L., Sappa, A.D., Vintimilla, B.X., Hammoud, R.I.:Deep learning based single image dehazing. In: 2018 IEEE/CVF Conference on Computer Vision and Pattern Recognition Workshops (CVPRW), pp. 1250–12507 (2018)
19. Bhaavan Sri Sailesh, A.:Arduino based smart street light system. In: 2021 3rd International Conference on Advances in Computing, Communication Control and Networking (ICAC3N), pp. 657–660 (2021)
20. Venkateswara Rao, Ch., Pathi, M.V.A., Sailesh, B.S.A:Arduino based electronic voting system with biometric and GSM features. In: 2022 4th International Conference on Smart Systems and Inventive Technology (ICSSIT), pp. 685–688 (2022)
21. Chand, R.P., Sri, V.B., Lakshmi, P.M.. Chakravathi, S.S., Veerendra, O.D.M., Rao, C.V.: Arduino based smart dustbin for waste management during Covid-19. In: 2021 5th International Conference on Electronics, Communication and Aerospace Technology (ICECA), pp. 492–496 (2021)
22. Shaik, A.R., Chandra, K.R., Raju, B.E., Budumuru, P.R.: Glaucoma Identification based on Segmentation and Fusion Technique. In: 2021 International Conference on Advances in Computing, Communication, and Control (ICAC3), pp. 1–4 (2021)

23. Budumuru, P.R., Kumar, G.P., Raju, B.E.:Hiding an image in an audio file using LSB audio technique. In: 2021 International Conference on Computer Communication and Informatics (ICCCI), pp. 1–4 (2021)
24. Salazar-Colores, S., Cruz-Aceves, I.: Single image dehazing using a multilayer perceptron. J. Electron. Imaging **27**, 1 (2018)
25. Yuan, K., Wei, J., Lu, W., Xiong, N.: Single Image Dehazing via NIN-DehazeNet. IEEE Access **7**, 181348–181356 (2019)
26. Li, J., Li, G., Fan, H.: Image dehazing using residual-based deep CNN. IEEE Access **6**, 26831–26842 (2018)

Performance Evaluation of Multiwavelet Transform for Single Image Dehazing

M. Ravi Sankar[1], P. Srinivas[1], V. Praveena[3], D. Bhavani[2], M. Sri Uma Suseela[3], Y. Srinivas[3], and Ch. Venkateswara Rao[3](✉)

[1] Department of ECE, Sasi Institute of Technology and Engineering, Tadepalligudem, India
[2] Department of ECE, SRKR Engineering College, Bhimavaram 534202, India
[3] Department of ECE, Vishnu Institute of Technology, Bhimavaram 534202, India
venkateswararao.c@vishnu.edu.in

Abstract. Images captured in poor lighting conditions (such haze, fog, mist, or smog) have a lower level of visibility because air particles deflect light. Single picture dehazing techniques can restore clarity to a single hazy image. Even though Dark channel prior (DCP) has been the most used method for image dehazing algorithms, it has poor picture quality and requires a lot of adjustments for real time applications such as; computer vision, object detection. However, previous studies such as; Dark Channel Prior (DCP) on image dehazing agonises from a gigantic processing time and sleaze of the original image. In addition, the dimensions of the images have a significant effect on the performance of the dehazing algorithms. Hence, the main motive of this work is to minimize the computational overload either by reducing the size of the image or by accurate transmission map estimation without compromising the image quality. Hence, this work is focused on reducing the computational time for image dehazing through multilevel discrete (Haar) wavelet transform, in which the image dimensions has been reduced without degrading quality. The performance measures of proposed algorithm have been analysed in terms of PSNR, MMSE and SSIM and compared with existing DCP algorithm. The simulation results proven that, the computational time for proposed algorithm has been reduced by 90% when compared to DCP without degrading image quality.

Keywords: Computational Time · Computer Vision · Dark Channel Prior · Dehazing · Image Dimension · Image Quality · Object Detection · Transmission map · Visibility · Wavelet Transform

1 Introduction

The pictures taken in outdoor sceneries frequently have haze, fog, or other atmospheric deterioration owing to medium particles that absorb and scatter light as it passages towards spectator. Due to this process, the quality will severely impair and which leads to immoral image quality. In contrast, a cloudy image is always detrimental to perception applications like tracking objects, observing the environment etc. To improve the image

N. Gupta et al. (Eds.): IC4S 2022, LNICST 472, pp. 125–133, 2023.
https://doi.org/10.1007/978-3-031-28975-0_10

quality for visualization and analysis as well as retrieving meaningful information from contaminated images, it is very crucial to remove the negative element. The removal of haze from photographs can be accomplished using a variety of techniques [1].

The optics of a computer vision framework are for the most part planned utilizing the presumption of splendid weather patterns, where the shading power of every pixel is exclusively related with the brilliance of the first scene. Consequently, learns at a beginning phase of computer vision errands specifically overlooked the state of awful climate [2]. Nonetheless, specialists soon understood the significance of picture reclamation procedures. Outside pictures are unavoidably antagonistically affected by the circumstances, and refraction, scattering and maintenance occur indeed, even on moderately sunny mornings, bringing about loss of itemized data and low differentiation. These debased info pictures lead to unfavorable effects on independent frameworks [3]. The model of haze formation has been depicted in Fig. 1.

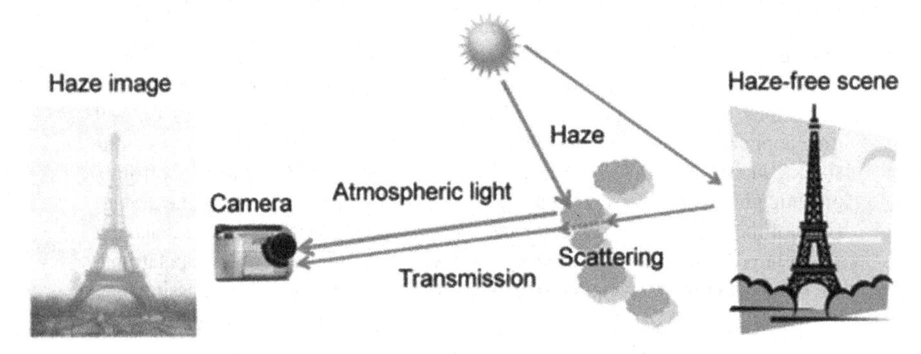

Fig. 1. Haze creation model [4].

Image dehazing with multiwavelet transform has a number of benefits over competing techniques [5]. The regularization elements of the optimization formulation can therefore enable flexibility in computing the dehazing solution by using a priori knowledge of the image and the atmospheric light transmission distribution, directing the computation to a desirable conclusion. Due to these benefits, high-quality dehazing performance can be achieved in a short amount of time. Hence, this work focuses on developing an algorithm for dehazing of images with high accuracy and less computational time by maintains adequate picture quality [6, 7].

2 Related Work

Most single picture dehazing calculations utilize either prior based picture handling determined to anticipate the broadcast, while deducing the barometrical light by exact techniques. A critical headway has the disclosure in [8], one channel that has an immaterial power among concealing coordinates in non-sky areas. The proposed algorithm in [9] offers the potential gains of straight inconvenience and feasibility yet moreover has drawbacks in flexibility, speed, and edge assurance, impelling numerous endeavors to manage the essential method.

The authors in [10] have consolidated the dehazing methodologies to move along the computational time. Thusly, two or three papers [11–14] credited the long dealing with season of the essential dehazing procedure to the delicate matting development, supplanting it with different channels. More of late, morphological redirection has partaken in the dehazing collaboration, vanquishing the absence of detail accomplished by the base channel [15]. The authors in [16] presented a joint improvement between equipment and calculation, accomplishing ongoing handling with restricted corruption. A new report [17] utilized a book unaided learning strategy by means of minimization of the computational time, taking care of the organization with certifiable cloudy pictures as opposed to ordinarily utilized engineered information to keep away from the conceivable space shift. In [18], the authors have presented an algorithm, accepting the brilliance and immersion are significantly different by the fog focus. In [19], a non-neighborhood dehazing (NLD) calculation for single pictures has been proposed. As indicated by their perceptions, the quantity of tones is a lot more modest than the quantity of pixels in a picture.

All through the advancement of knowledge-based dehazing approaches, the longing for elite execution has driven the advancement of progressively complex procedures, for example, multi-scale designs. This pattern has decreased the commonsense utilization of single picture dehazing algorithms, specifically for asset obliged applications where ongoing activity with restricted computational assets is a key driver. Numerous new investigations based on Arduino integrated with IOT [20–24] perceive the requirement for ongoing handling and the advancement of methods to lessen the capacity, intricacy, handling time and other related angles without compromising the presentation. The proposed image dehazing algorithm mainly focused on reducing the dimensions of the image through multiwavelet transform there by achieving the less computational time when compared to DCP algorithm by maintain bearable picture quality.

3 Methodology

The primary disadvantage in DCP method is its computational time. Pragmatic applications require less opportunity to compute the dehazed picture from haze pictures. To beat this downside a superior adaptation of DCP calculation which is a Fast DCP has been proposed for dehazing of images. Image dehazing based on DCP method [25] in which the transmission medium $t'(x)$ is determined by utilizing confined area. In this way, the border area has been unbending for $t'(x)$. To smoothen these edges the algorithm utilized transmission map refinement, which is a tedious interaction and does not appropriate for pragmatic applications. The haze image is a combination of image coordinates, observed image, haze free image, atmospheric light, scattering coefficient and scene depth [24], which is represented using (1). The dark channel from the input hazy image is estimated by the majority of traditional DCP-based dehazing techniques and it is represented using (2).

$$I_{haze}(x) = J_{haze-free}(x)e^{-\beta d(x)} + A\left(1 - e^{-\beta d(x)}\right) \quad (1)$$

$$J^{dark}(x) = \min_{y \in \Omega(x)}\left(\min_{c \in \{r,g,b\}} J^{c}(y)\right) \quad (2)$$

3.1 Image Model

The hazed image has two additive components. The first element, or clear image, is made up of light that has been reflected off of an object's surface. The scattering transmission, or haze, is the second component. The optical model of an image can be expressed using (3).

$$I_c = J_c \odot t + a_c(1 - t), c = 1, 2, 3\ldots \quad (3)$$

3.2 Sub-band Image Model

Using DHWT to quickly break down the image model into a bank of frequency sub bands based on the low pass characteristic of the light broadcast dispersal. Thus, the information about t is already included in the decomposed picture model in the low-frequency region and may be used to estimate it. The computing complexity of the dehazing process can be significantly decreased by using this sub-band image model with a smaller size. For example, t can be characterized by means of (4) when the atmosphere is homogeneous.

$$t = e^{-\beta d} \quad (4)$$

Based on discrete haar wavelet transform, the sub-band image model can be expressed using (5) based on the approximate 2-D matrix dimension W.

$$\mathrm{W}_{\mathrm{p,q}}^{\mathrm{n}} = {}^{1}\!\big/\sqrt{2^{\mathrm{k-p}}} \text{ if } \mathrm{q} \times 2^{\mathrm{k-p}} \le \mathrm{n} < \left(\mathrm{q} + \frac{1}{2}\right) \times 2^{\mathrm{k-p}} \quad (5)$$

4 Simulation Results

4.1 Subjective Analysis

The simulation has been performed using MATLAB with 1.5 GHz operating frequency. Firstly, the optical model of the input hazy image has been estimated. The subjective analysis has been carried out for various hazy images (see Fig. 2).

Fig. 2. Simulation results of proposed dehazing algorithm for various haze images.

After estimating the optical model of an image, the image transmission distribution has been estimated. By keeping both the values of scene depth d and β are positive, the transmission distribution t has been evaluated. Then, the sub-band image model has been created with reduced dimension of the image for achieving less computational time. Subjective analysis is the analysis which uses the observers to make the quality estimation based on their visual opinion of the image.

4.2 Objective Analysis

The objective analysis has been conducted for the proposed methodology in terms of PSNR, MSE and computational time. The detailed analysis has been depicted in Table 1. The PSNR and mean square error has been evaluated using Eqs. (6) and (7) respectively.

$$PSNR = 20\log(\frac{Max}{\sqrt{MSE}}) \tag{6}$$

$$MSE = \frac{1}{MM}\sum_{i=0}^{m-1}\sum_{j=0}^{n-1}[I(i,j) - K(i,j)]^2 \tag{7}$$

Table 1. Comparison of MSE, PSNR and SSIM values of DCP and DHWT algorithms for various images

Image name	Parameter	DCP	DHWT
Image-1	MSE	0.0087	0.0073
	PSNR	62.7544	69.5266
	SSIM	0.8341	0.9654
Image-2	MSE	0.0065	0.0039
	PSNR	68.4634	72.2325
	SSIM	0.8535	0.9689
Image-3	MSE	0.0079	0.0098
	PSNR	66.7544	68.2286
	SSIM	0.7986	0.9316
Image-4	MSE	0.0092	0.0083
	PSNR	67.8243	68.9305
	SSIM	0.8742	0.8917
Image-5	MSE	0.0079	0.0069
	PSNR	63.4570	69.7550
	SSIM	0.8721	0.9649
Image-6	MSE	0.0069	0.0068

(continued)

Table 1. (*continued*)

Image name	Parameter	DCP	DHWT
	PSNR	69.9675	69.8259
	SSIM	0.9436	0.9368

Along with the objective analysis; the computational complexity has been estimated for both the algorithms (see Table 2).

Table 2. Comparison of computational time.

Image name	DCP (s)	Proposed (s)
Image-1	9.874	0.6688
Iamge-2	9.774	0.3388
Image-3	9.869	0.5840
Image-4	10.057	0.8681
Image-5	9.877	1.4269
Image-6	8.7675	0.4528

5 Conclusion

Image dehazing expulsion techniques have become more valuable for many picture handling and computer vision applications. All the dehazing techniques helpful for reconnaissance, for remote detecting and submerged imaging, photography and so forth A large portion of the techniques depend on the assessment of climatic light and transmission map. In, this work a DHWT for dehazing the images have been proposed. The proposed method performs the dehazing without degrading the picture quality and with less computational time when compared to DCP for single image dehazing.

References

1. Kermani, E., Asemani, D.: A robust adaptive algorithm of moving object detection for video surveillance. EURASIP J. Image Video Process. **2014**(1), 1–9 (2014)
2. Ozaki, M., Kakimuma, K., Hashimoto, M., Takahashi, K.: Laser-based pedestrian tracking in outdoor environments by multiple mobile robots. In: Proceedings of Annual Conference on IEEE Industrial Electronics Society, pp. 197–202. IECON, Melbourne (2011)
3. Lee, S., Yun, S., Nam, J.-H., Won, C.S., Jung, S.-W.: A review on dark channel prior based image dehazing algorithms. EURASIP J. Image Video Process. **2016**(1), 1–23 (2016)
4. Anwar, M.I., Khosla, A.: Vision enhancement through single image fog removal. Eng. Sci. Technol., Int. J. **20**(3), 1075–1083 (2017)

5. Badhe, M.V., Prabhakar, L.R.: A survey on haze removal using image visibility restoration technique. Int. J. Comput. Sci. Mob. Comput. **5**(2), 96–101 (2016)
6. Cai, B., Xu, X., Jia, K., Qing, C., Tao, D.: Dehazenet: an end-to-end system for single image haze removal. IEEE Trans. Image Process. **25**(11), 5187–5198 (2016)
7. Gibson, K.B., Vo, D.T., Nguyen, T.Q.: An investigation of dehazing effects on image and video coding. IEEE Trans. Image Proces. **21**(2), 662–673 (2012). https://doi.org/10.1109/TIP.2011.2166968
8. He, K., Sun, J., Tang, X.: Single image haze removal using dark channel prior. Pattern Anal. Mach. Intell. IEEE Trans. **33**(12), 2341–2353 (2011)
9. Gibson, K.B., Nguyen, T.Q.: On the efectiveness of the Dark Channel Prior for single image dehazing by approximating with minimum volume ellipsoids. In: IEEE international conference on acoustics, speech, and signal processing (ICASSP), pp. 1253–1256 (2011)
10. Xie, B., Guo, F., Cai, Z.: Improved single image dehazing using dark channel prior and multi-scale retinex. In: Proceedings of international conference on intelligent system design and engineering application. IDEA 2010, pp 848–851 (2018)
11. Park, D., Han, D.K., Ko, H.: Single image haze removal with WLS-based edge-preserving smoothing flter. In: IEEE International conference on acoustics, speech and signal processing (ICASSP), pp. 2469–2473 (2013)
12. Gibson, K.B., Nguyen, T.Q.: Fast single image fog removal using the adaptive Wiener flter. In: IEEE international conference on image processings, ICIP, pp. 714–718 (2013)
13. Xiao, C., Gan, J.: Fast image dehazing using guided joint bilateral flter. Vis. Comput. **28**, 713–721 (2012)
14. Zhang, Q., Li, X.: Fast image dehazing using guided flter. In: IEEE 16th international conference on communication technology (ICCT), pp. 182–185 (2015)
15. Salazar-Colores, S., et al.: A fast image dehazing algorithm using morphological reconstruction. IEEE Trans. Image Proc. **28**(5), 2357–2366 (2019)
16. Lu, J., Dong, C.: DSP-based image real-time dehazing optimization for improved dark-channel prior algorithm. J. Real-Time Image Proc. **17**(5), 1675–1684 (2019)
17. Golts, A., Freedman, D., Elad, M.: Unsupervised single image dehazing using dark channel prior loss. IEEE Trans. Image Proc. **29**, 2692–2701 (2020)
18. Zhu, Q., Mai, J., Shao, L.: A fast single image haze removal algorithm using color attenuation prior. Image Process. IEEE Trans. **24**(11), 3522–3533 (2015)
19. Berman, D., Treibitz, T., Avidan, S.: Non-local image dehazing. In: Computer vision and pattern recognition (CVPR), pp. 1674–1682 (2016)
20. Bhaavan Sri Sailesh, A., Sudha Madhavi, A., Venkata Pavan, G., Sravanthi, I., Karthik Sai Kiran, B., Venkateswara Rao, Ch.: Arduino based smart street light system. In: 2021 3rd International Conference on Advances in Computing, Communication Control and Networking (ICAC3N), pp. 657–660 (2021)
21. Venkateswara Rao, Ch., Pathi, M.V.A., Sailesh, B.S.A.: Arduino based electronic voting system with biometric and GSM features. In: 2022 4th International Conference on Smart Systems and Inventive Technology (ICSSIT), pp. 685–688 (2022)
22. Chand, R.P., Sri, V.B., Lakshmi, P.M., Chakravathi, S.S., Veerendra, O.D.M., Rao, C.V.: Arduino based smart dustbin for waste management during Covid-19. In: 2021 5th International Conference on Electronics, Communication and Aerospace Technology (ICECA), pp. 492–496 (2022)
23. Shaik, A.R., Chandra, K.R., Raju, B.E., Budumuru, P.R.: Glaucoma identification based on segmentation and fusion techniques. In: 2021 International Conference on Advances in Computing, Communication, and Control (ICAC3), pp. 1–4 (2021)

24. Budumuru, P.R., Kumar, G.P., Raju, B.E.: Hiding an image in an audio file using LSB audio technique. In: 2021 International Conference on Computer Communication and Informatics (ICCCI), pp. 1–4 (2021)
25. Iwamoto, Y., Hashimoto, N., Chen, Y.-W.: Fast dark channel prior based haze removal from a single image. In: 2018 14th International Conference on Natural Computation, Fuzzy Systems and Knowledge Discovery (ICNC-FSKD), pp. 458–461 (2018)

Performance Evaluation of Fast DCP Algorithm for Single Image Dehazing

I. Sravanthi, A. Sudha Madhavi, G. Venkata Pavan, B. Karthik Sai Kiran, A. Bhaavan Sri Sailesh, Ch. V V Santhosh Kumar, and Ch. Venkateswara Rao(✉)

Electronics and Communication Engineering, Vishnu Institute of Technology, Bhimavaram 534202, Andhra Pradesh, India
venkateswararao.c@vishnu.edu.in

Abstract. Poor weather circumstances including fog, dust, and atmospheric dispersion of other particles can degrade the clarity of photos taken in outdoor settings. High-level computer vision tasks like picture segmentation and object detection may become more difficult as a result of this issue. However, earlier research on image dehazing, such as Dark Channel Prior (DCP), suffers from a significant computing burden and degradation of the original image, such as oversaturation and halos. In this work, an improved image dehazing approach (Fast DCP) based on down sampling has been proposed. The proposed algorithm executes the dehazing of a haze image with a smaller amount computational time without degrading the quality of the image. The performance measures of proposed algorithm have been analysed in terms of PSNR, MMSE and SSIM and compared with existing DCP algorithm. The simulation results proven that, the computational time for proposed algorithm has been reduced by 70% when compared to DCP without degrading image quality.

Keywords: Computer Vision · Computational Time · Dark Channel Prior · Dehazing · Object Detection · Segmentation · Scattering

1 Introduction

Haze is a typical environmental peculiarity brought about by tiny samples such as; dirt, fume, fog in the air, driving to debasement of picture lucidity. Customarily, specialists treated picture dehazing as a picture handling method to recuperate image subtleties and this confined the utilization of dehazing calculations to a restricted scope of uses. Be that as it may, the quick improvement of independent frameworks, artificial insight, and the prerequisites of significant level computer vision assignments has prompted re-established examination into further developed picture dehazing procedures [1]. The improvement of dehazing estimations has, therefore, become perhaps the best area of artificial intelligence and an undeniably basic supplication is the utilization of dehazing calculations in order to move along the showcase of the self-governed structures and stages for antagonistic barometrical circumstances.

N. Gupta et al. (Eds.): IC4S 2022, LNICST 472, pp. 134–143, 2023.
https://doi.org/10.1007/978-3-031-28975-0_11

The optics of a computer vision framework are for the most part planned utilizing the presumption of splendid weather patterns, where the shading power of every pixel is exclusively related with the brilliance of the first scene. Consequently, learns at a beginning phase of computer vision errands specifically overlooked the state of awful climate [2]. Nonetheless, specialists soon understood the significance of picture reclamation procedures. Outside pictures are unavoidably antagonistically affected by the circumstances, and refraction, scattering and maintenance occur indeed, even on moderately sunny mornings, bringing about loss of itemized data and low differentiation. These debased info pictures lead to unfavorable effects on independent frameworks [3]. The model of haze formation has been depicted in Fig. 1.

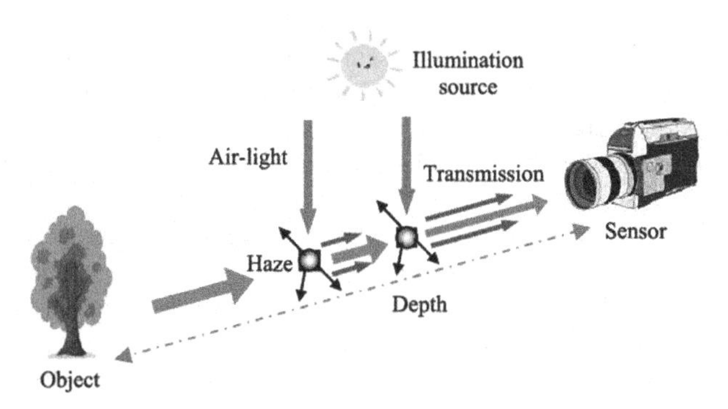

Fig. 1. Model of haze creation [3]

Haze expulsion is exceptionally alluring in various fields like computer vision calculations, picture handling and photography [4, 5]. To begin with, the amputation of haze rises the perceptibility of a dim sceneries formed by air particles. The relating picture is the scene Luminant in most computer picture handling approaches, from higher scale picture processing to advanced scale shape recognition. The assessment of these methods relies upon the scenes. In the event that the picture or scene is dull, vision calculations face many issues and try not to show proficient execution. In this way, eliminating fog is required for better outcomes and productivity. The awful pictures can be put to all the more likely use. Scattering relay on the scene length, and this sleaze is called three-dimensional variation. Expulsion of murkiness from the picture expands the perceivability of dim picture brought about by the air particles. Hence, this work focuses on developing an algorithm for dehazing of images with high accuracy and less computational time by maintains adequate picture quality [6, 7].

2 Literature Review

Most single picture dehazing calculations utilize either prior based picture handling determined to anticipate the broadcast, while deducing the barometrical light by exact techniques. A crucial headway has the disclosure in [8], which elaborates the image dehazing phenomenon. The proposed algorithm in [9] offers the potential gains of straight inconvenience and feasibility yet moreover has drawbacks in flexibility, speed, and edge assurance, impelling numerous endeavours to manage the essential method.

The authors in [10] have consolidated the dehazing methodologies to move along the computational time. Thusly, two or three papers [11–14] credited the long dealing with season of the essential dehazing procedure to the delicate matting development, supplanting it with different channels. More of late, morphological redirection has partaken in the dehazing collaboration, vanquishing the absence of detail accomplished by the base channel [15]. The authors in [16] presented a joint improvement between equipment and calculation, accomplishing ongoing handling with restricted corruption. A new report [17] utilized a book unaided learning strategy by means of minimization of the computational time, taking care of the organization with certifiable cloudy pictures as opposed to ordinarily utilized engineered information to keep away from the conceivable space shift. In [18], the authors have presented an algorithm, accepting the brilliance and immersion are significantly different by the fog focus. In [19], a non-neighborhood dehazing (NLD) calculation for single pictures has been proposed. As indicated by their perceptions, the quantity of tones is a lot more modest than the quantity of pixels in a picture.

All through the advancement of knowledge-based dehazing approaches, the longing for elite execution has driven the advancement of progressively complex procedures, for example, multi-scale designs. This pattern has decreased the commonsense utilization of single picture dehazing algorithms, specifically for asset obliged applications where ongoing activity with restricted computational assets is a key driver. Numerous new investigations based on Arduino integrated with IOT [20–24] perceive the requirement for ongoing handling and the advancement of methods to lessen the capacity, intricacy, handling time and other related angles without compromising the presentation. The proposed Fast DCP algorithm mainly focused on single image dehazing with less computational time when compared to DCP algorithm by maintain bearable picture quality.

3 Methodology

The primary disadvantage in DCP method is its computational time. Pragmatic applications require less opportunity to compute the dehazed picture from haze pictures. To beat this downside a superior adaptation of DCP calculation which is a Fast DCP has been proposed for dehazing of images. Image dehazing based on DCP method [25] in which the transmission medium $t'(x)$ is determined by utilizing confined area. In this way, the border area has been unbending for $t'(x)$. To smoothen these edges the algorithm utilized transmission map refinement, which is a tedious interaction and does not appropriate for pragmatic applications. Hence, to improve the computational time, a Fast DCP method has been proposed. The proposed algorithm consists of three basic steps in working such as;

- Down-sampling
- Estimation of pixel-wise approximation
- Assessment of robust ambient light

3.1 Down-Sampling Based Computation

The computational time is solidly comparing to picture size. By and large, the dim channel picture has low spatial repeat. There are no tremendous ramifications for the evaluated dim channel regard in any event, when down-inspected picture and a short time later add it to special size ensuing to determining the dim channel picture and including light. The picture is down-inspected to *N/4, M/4* utilizing the mean in this technique.

3.2 Assessment Dark Channel Image Through Element-Wise

The proposed calculation assesses the dark channel esteem utilizing one pixel rather than 15 × 15 pixel to lessen the computational time without trailing the spatial data like edge. The DCP utilizes pixel (1 × 1 fix) for estimation of dim channels, there is no requirement for delicate matting interaction to smoothen the created picture, which is a tedious cycle. Notwithstanding this the proposed technique utilizes down-tested picture and one pixel for estimation of dull channel which diminish the computational time essentially [25]. Allow us to think about the worth of direct light J in obscurity channel as nothing, therefore the right-hand side's starting term in Eq. (1) can be omitted.

In the proposed technique, the worth of dim divert in noticed light I determined by 1 × 1 fix is bigger than the ordinary dim channel esteem determined by 15 × 15 fix (see Eq. (2)). Because of the expulsion of spatial limiting, there an opportunity that the dim divert esteem in straight light J has little worth (see Eq. (3)).

$$\min_{y\in\Omega(x)}\left(\min_{c}\left(\frac{I^c(y)}{A^c}\right)\right)=\tilde{t}(x)\min_{c}\left(\min_{c}\frac{j^c(y)}{A^c}\right)+(1-\tilde{t}(x)) \tag{1}$$

$$T_J(x)=1-\gamma\frac{\left(\min_{c\in\{R,G,B\}}\left(\frac{I^c(x)}{A^c}\right)-\min_{x\in\Omega}\left(\min_{c\in\{R,G,B\}}\left(\frac{I^c(x)}{A^c}\right)\right)\right)}{\left(\max_{x\in\Omega}\left(\min_{c\in\{R,G,B\}}\left(\frac{I^c(x)}{A^c}\right)\right)-\min_{x\in\Omega}\left(\min_{c\in\{R,G,B\}}\left(\frac{I^c(x)}{A^c}\right)\right)\right)} \tag{2}$$

$$t'(x)=\frac{\left(1-\omega\min_{y\in\Omega(x)}\left(\min_{c\in\{R,G,B\}}\frac{I^c(y)}{A^c}\right)\right)}{T_J(x)} \tag{3}$$

3.3 Estimation of Robust Ambient Light

In DCP strategy [25], at first the creators have picked the top 0.1% most brilliant pixels in obscurity channel of cloudiness picture I, and from these pixels the most noteworthy power pixels has been recognized. Prior techniques for dehazing, overlooks the little white article because of dim channel picture which is assessed by 15 × 15 fix. In any case, in the Fast DCP the dull channel is determined by utilizing 1 × 1 fix. What is more, the DCP technique needs an opportunity to sort the top 0.1% most splendid pixels in obscurity channel of fog picture *I*.

4 Simulation Analysis

The simulation has been performed using MATLAB with 1.5 GHz operating frequency. Firstly, the input haze image has been subjected for proposed algorithm (see Fig. 2(a)). First, it is down sampled by four and applied for dark channel construction (see Fig. 2(b)).

Fig. 2. (a) Input Hazy Image, (b) Down-Sampled Image, (c) Dark Channel Image, (d) Transmission Map, (e) Dehazed image.

Fig. 3. (a) Input Hazy Image, (b) Dehazed Image using DCP algorithm, (c) Proposed

The dark channel formation involved in many phases with respect to DCP and the output mage padded with zeros (see Fig. 2(c)).

After the dark channel construction, transmission map has been constructed (see Fig. 2(d)). Finally, the dehazed image has been estimated (see Fig. 2(e)). The proposed algorithm has been justified by performing subjective and objective analysis respectively. The performance measures have been evaluated for various hazy images.

4.1 Subjective Analysis

Subjective analysis is the analysis which uses the observers to make the quality estimation based on their visual opinion of the image. The subjective analysis has been carried out for various hazy images (see Fig. 3) such as; Trees and Trunks, Colorful flowers, Dolls, Village and for an Underwater image. The subjective analysis has been carried out by performing the down sampling of an input hazy image there by constructing dark channel image and transmission map. Finally, the dehazed image has been obtained after transmission map. The subjective analysis has been carried out for both existing DCP and proposed Fast DCP algorithms. Figure 3 depicted that the dehazed images corresponding to both DCP and Proposed Fast DCP algorithms. The input hazy images have been taken from the existing data base of hazy images.

4.2 Objective Analysis

The objective analysis has been conducted for the proposed methodology in terms of PSNR, MSE and computational time. The detailed analysis has been depicted in Table 1. Along with the objective analysis; the computational complexity has been estimated for both the algorithms (see Table 2). The PSNR and mean square error has been evaluated for proposed Fast DCP algorithm and compared with existing DCP algorithm. The PSNR and MSE have been represented using (4) and (5) respectively. In addition to PSNR and MSE, the structural similarity index measurement (SSIM) has been evaluated for both proposed Fast DCP algorithm and existing DCP algorithm in order to know the quality of dehazed images.

$$PSNR = 20\log\left(\frac{Max}{\sqrt{MSE}}\right) \tag{4}$$

$$MSE = \frac{1}{mn}\sum_{i=0}^{m-1}\sum_{j=0}^{n-1}[I(i,j) - K(i,j)]^2 \tag{5}$$

Table 1. Comparison of MSE, PSNR and SSIM values of DCP and fast DCP algorithms for various images

Image name	Parameter	DCP	Proposed
Trees and trunks	MSE	0.0087	0.0069
	PSNR	62.7544	69.7556
	SSIM	0.8341	0.9689
Colourful flowers	MSE	0.0065	0.0073
	PSNR	68.4634	67.1674
	SSIM	0.8535	0.8425
Dolls	MSE	0.0079	0.0067
	PSNR	66.7544	69.8670
	SSIM	0.7986	0.8466
Village	MSE	0.0092	0.0086
	PSNR	67.8243	69.5269
	SSIM	0.8742	0.9368
Under water	MSE	0.0079	0.0059
	PSNR	63.4570	69.8249
	SSIM	0.8721	0.9659

Table 2. Comparison of computational time.

Image name	DCP (s)	Proposed (s)
Trees and trunks	9.874	2.281
Colourful flowers	9.774	2.148
Dolls	9.869	2.119
Village	10.057	2.068
Under water	9.877	2.527

5 Conclusion

Image dehazing expulsion techniques have become more valuable for many picture handling and computer vision applications. All the dehazing techniques helpful for reconnaissance, for remote detecting and submerged imaging, photography and so forth A large portion of the techniques depend on the assessment of climatic light and transmission map. In, this work a Fast DCP method for dehazing the images have been proposed. The proposed method performs the dehazing without degrading the picture quality and with less computational time when compared to DCP for single image dehazing. It is clearly depicted from results, the computational time with proposed methodology has been reduced by 70% when compared to DCP algorithm. Image quality has achieved a value of 69.8670 (PSNR) with proposed algorithm whereas, for DCP the PSNR value is 66.7544.

References

1. Kermani, E., Asemani, D.: A robust adaptive algorithm of moving object detection for video surveillance. EURASIP J. Image Video Process. **2014**(1), 1–9 (2014). https://doi.org/10.1186/1687-5281-2014-27
2. Ozaki, M., Kakimuma, K., Hashimoto, M., Takahashi, K.: Laser-based pedestrian tracking in outdoor environments by multiple mobile robots. In: Proceedings of Annual Conference on IEEE Industrial Electronics Society, pp. 197–202. IECON, Melbourne (2011)
3. Lee, S., Yun, S., Nam, J.-H., Won, C.S., Jung, S.-W.: A review on dark channel prior based image dehazing algorithms. EURASIP J. Image Video Process. **2016**(1), 1–23 (2016). https://doi.org/10.1186/s13640-016-0104-y
4. Anwar, M.I., Khosla, A.: Vision enhancement through single image fog removal. Eng. Sci. Technol. Int. J. **20**(3), 1075–1083 (2017)
5. Badhe, M.V., Prabhakar, L.R.: A survey on haze removal using image visibility restoration technique. Int. J. Comput. Sci. Mob. Comput. **5**(2), 96–101 (2016)
6. Cai, B., et al.: Dehazenet: an end-to-end system for single image haze removal. IEEE Trans. Image Process. **25**(11), 5187–5198 (2016)
7. Gibson, K.B., et al.: An investigation of dehazing effects on image and video coding. IEEE Trans. Image Process. **21**(2), 662–673 (2012)
8. He, K., Sun, J., Tang, X.: Single image haze removal using dark channel prior. Pattern Anal. Mach. Intell. IEEE Trans. **33**(12), 2341–2353 (2011)
9. Gibson, K.B., Nguyen, T.Q.: On the effectiveness of the dark channel Prior for single image dehazing by approximating with minimum volume ellipsoids. In: IEEE International Conference on Acoustics, Speech, and Signal Processing (ICASSP), pp. 1253–1256 (2011)
10. Xie, B., Guo, F., Cai, Z.: Improved single image dehazing using dark channel prior and multi-scale retinex. In: Proceedings of International Conference on Intelligent System Design and Engineering Application, pp. 848–851. IDEA 2010 (2018)
11. Park, D., Han, D.K., Ko, H.: Single image haze removal with WLS-based edge-preserving smoothing filter. In: IEEE International Conference on Acoustics, Speech and Signal Processing (ICASSP), pp. 2469–2473 (2013)
12. Gibson, K.B., Nguyen, T.Q.: Fast single image fog removal using the adaptive Wiener filter. In: IEEE International Conference on Image Processing, pp. 714–718. ICIP (2013)
13. Xiao, C., Gan, J.: Fast image dehazing using guided joint bilateral filter. Vis. Comput. **28**, 713–721 (2012)
14. Zhang, Q., Li, X.: Fast image dehazing using guided filter. In: IEEE 16th International Conference on Communication Technology (ICCT), pp. 182–185 (2015)
15. Salazar-Colores, S., et al.: A fast image dehazing algorithm using morphological reconstruction. IEEE Trans. Image Proc. **28**(5), 2357–2366 (2019)
16. Lu, J., Dong, C.: DSP-based image real-time dehazing optimization for improved dark-channel prior algorithm. J. Real Time Image Proc. **17**(5), 1675–1684 (2019). https://doi.org/10.1007/s11554-019-00933-3
17. Golts, A., Freedman, D., Elad, M.: Unsupervised single image dehazing using dark channel prior loss. IEEE Trans. Image Proc. **29**, 2692–2701 (2020)
18. Zhu, Q., Mai, J., Shao, L.: A fast single image haze removal algorithm using color attenuation prior. Image Process. IEEE Trans. **24**(11), 3522–3533 (2015)

19. Berman, D., Treibitz, T., Avidan, S.: Non-local image dehazing. In: Computer Vision and Pattern Recognition (CVPR), pp. 1674–1682 (2016)
20. Bhavan Sri Sailesh, A., Sudha Madhavi, A., Venkarta Pavan, G., Sravanthi, I., Karthick Sai Kiran, B., Venkateshwara Rao, Ch.: Arduino based smart street light system. In: 2021 3rd International Conference on Advances in Computing, Communication Control and Networking (ICAC3N), pp. 657–660 (2021)
21. Venkateshwara Rao, Ch., Pathi, A.M.V., and Sailesh, A.B.S.: Arduino based electronic voting system with biometric and GSM features. In: 2022 4th International Conference on Smart Systems and Inventive Technology (ICSSIT), pp. 685–688 (2022)
22. Chand, R.P., Sri, V.B., Lakshmi, P.M., Chakravathi, S.S., Veerendra, O.D.M., Rao, C.V.: Arduino based smart dustbin for waste management during COVID-19. In: 5th International Conference on Electronics, Communication and Aerospace Technology (ICECA), pp. 492–496 (2022)
23. Shaik, A.R., Chandra, K.R., Raju, B.E., Budumuru, P.R.: Glaucoma identification based on segmentation and fusion techniques. In: 2021 International Conference on Advances in Computing, Communication, and Control (ICAC3), pp. 1–4 (2021)
24. Budumuru, P.R., Kumar, G.P., Raju, B.E.: Hiding an image in an audio file using LSB audio technique. In: 2021 International Conference on Computer Communication and Informatics (ICCCI), pp. 1–4 (2021)
25. Iwamoto, Y., Hashimoto, N., Chen, Y.-W.: Fast dark channel prior based haze removal from a single image. In: 2018 14th International Conference on Natural Computation, Fuzzy Systems and Knowledge Discovery (ICNC-FSKD), pp. 458–461 (2018)

Non Destructive Analysis of Crack Using Image Processing, Ultrasonic and IRT: A Critical Review and Analysis

P. Ramani[1], V. Subbiah Bharathi[1], and S. Sugumaran[2(✉)]

[1] SRM Institute of Science and Technology, Ramapuram, Chennai, India
ramanip@srmist.edu.in, director@srmrmp.edu.in
[2] Vishnu Institute of Technology, Bhimavaram, AP, India
sugumaran.s@vishnu.edu.in

Abstract. Crack is one of the most important surfaces damage to monuments, concrete structures, buildings, and roads. Manually examining damage is time- and labor-intensive. Crack irregularity measurements are challenging and need more expertise. Thus, develop automatic crack detection using the image processing method. This article reveals the various strategies to distinguish the crack length, width, and depth utilizing different automatic crack detection methods. In this, 53 papers describe the detection of cracks and other decay measurements. The investigation is given on the survey and dependent on the Infrared Thermography method, Ultrasonic imaging, and Image processing. The main aim of this paper is to summarize and compare the few strategies used in various Non-Destructive Techniques. Detection of the crack using Deep Learning achieves with maximum accuracy of 98%. Finally, we represent different issues that can be valuable for inquiring about to achieve further investigation on this detection.

Keywords: Degradation · Infrared thermography · Morphological operation · Ultrasonic testing · Support vector machine

1 Introduction

India is a country with a very deep and significant history and a grand heritage. Its culture and traditions are very unique, varied and celebrated the world over. Each and every district in the country has its own set of heritage structures, traditions and cultural practices. Unfortunately the cultural heritage structures and road which are exposed to the environmental changes for thousands of years undergo severe damages. The road is decayed due to frequent accessing of vehicles with a heavy load and aging. The conservation of heritage structures and road necessitates assessment of weathering decay before planning the restoration activity. The conventional method of measuring the decay involves collecting the sample from the site and conducting lab tests. Those invasive techniques are not suitable for protected heritage structures. In this context, in situ non-destructive testing methods such as Image Processing methods/machine learning are most suitable for weathering decay assessment.

N. Gupta et al. (Eds.): IC4S 2022, LNICST 472, pp. 144–155, 2023.
https://doi.org/10.1007/978-3-031-28975-0_12

Monuments are decayed by the climate, man, rain, and other parameters. Our analysis shows how building structures like beams and concrete structures deteriorate and frequently develop cracks. Building structures are beams, concrete structures are decayed by continuous stress, and cyclic loading due to the crack is produced and cause material discontinuities. Early detection of such failures is necessary to save the building and historic cite structures. Many image processing detection methods were presented due to their ease of processing. These techniques can be divided into two groups: integrated algorithms and morphological approaches [1].

A detailed review of destructive and non-destructive techniques has been discussed in this following section. In a nutshell, it can be said that the survey is divided into four sections. Among them the first section is being dealt with the description of destructive and Non-destructive methods, the second one is about the review of various destructive and non-destructive methods and the comparison of corresponding measurement techniques, the third one is about image-based crack detection and recognition, the fourth one illustrate the accuracy based analysis, the fifth one describes the results and discussion and finally in the sixth section is about the conclusion and future work. Some of the destructive and Non-destructive methods are explained in this section.

1.1 Scanning Electron Microscopy

The images of a stone are created by continuously passing the beam of light. It provides the information of both micro and nanometric combination and also perform the analysis based on material composition [2].

1.2 Infrared Thermography

The thermal distribution of the surface is detected by the infrared method. It is used to find heat production and controls the temperature of a component. It can identify discontinuities/interfaces, defects, and voids and is difficult to detect the depth [3–5].

1.3 Ultrasonic Testing (UT)

Through the transducer, the high-frequency signal are passed to the rock. The flaws in material or parts are determined by the amount of transmitted and received energy. It is mainly used to identify the subsurface defects in the wood, plastics, and material and also difficult to inspect the thin parts. Depth is determined utilizing ultrasonic speed [6–12].

$$\text{Ultrasonic velocity} = \frac{\textit{Measuring distance}}{\textit{Delivery time}} \tag{1}$$

1.4 3Dimensional Terrestrial Laser Scanners

It is a new sensing technology, and it gives complex shape (like caves, historic sites). It can collect clouds of data points in few seconds and represented by horizontal and vertical view.

1.5 Image Processing

It is Non-destructive methods that processes the image automatically and obtain useful information. It uses the software to perform image processing operations on digital images to classify, extract and recognize pattern on surface [13–16]. It gives a detailed intervention compatibility and mapping of conservation.

2 Related Work

There are two techniques for surveying the decay of heritage structures and road, and they are Non-destructive and destructive methods [17]. The destructive methods are Transmitted Light Microscopy, X-ray diffraction, and Scanning Electron Microscopy [18]. In the said techniques, samples have been removed from the heritage structures and taken into the research center for examination which would eventually devastate a significant bit of monument structure. Technique of this type would not be prudent to survey the monuments.

Vibration assessment is said to be possible on a wide extent of structures such as bridges and monuments. But for these assessments the dynamic attributes of the structures like characteristic recurrence, methods of vibration, so on and so forth must be chosen delicately.

Crack has been identified and categorized by utilizing both Non-destructive and destructive techniques [1]. Some of the Non Destructive methods are Schmidt hammer that quantifies the strength of the stone, Ultrasonic imaging, Infrared Thermography and 3D Terrestrial Laser Scanner and image processing [19]. A brief description and comparison of these methods are given below and summarized in Table 1.

Antonia Moropoulou et al. (2013) proposed conservation methods for cultural heritage structures. They used various Non-destructive methods such as Digital Image Processing, Ultrasonic Testing, Infrared Thermography, and Fiber Optic Microscopy to study the decay of cultural heritage structures and the same had been validated in the lab. The ultrasonic testing technique had been used to recognize the outside and inside surface defects [20] in which a high-frequency signal had been used to identify the defects inside the stone. These techniques give flexible highlights of notable materials and furthermore, these techniques had been dealt in aiding the crack recognition and consideration of the structure. Here, the profundity of the influenced cover interior of the material is resolved [21]. An Infrared-Thermography (IRT) technique had been used to distinguish the imperfections of structural surface during night light and day light [22].

Brooks et al. (2018) developed a model to recognize the crack which has a thermal camera. The Camera had been used for distinguishing the impression of an infrared source from the exterior of the crack and the same had been used further to recognize the deformities in huge outside regions [16].

Christian Garnier et al. (2011) proposed some NDT methods and those methods were done on composite specimens to find the defect. The Graphical Visual Inspection had been used to identify the point and zone of the damaged parts. The consequences of the three strategies such as IRT, UT and Shearography are looked at and furthermore profundity is estimated using Ultrasonic Testing Method [23].

Non-destructive ultrasonic testing was used by Francesco et al. (2014) to evaluate the building materials of monuments, and elastic qualities were linked to the materials' physical and petrographic characteristics. The procedures include evaluation, observation of the weathering process, and use of refraction techniques to measure the thickness of the weather shallow area of the masonry [24]. An IRT strategy had been identified to recognize the delamination and debonding of the crack. To assess the measurement of a round and hollow specimen IRT had been utilized.

A novel technique for determining the depth of historic structures was developed by Pascale et al. (2015). The obtained result explained about the hardness of the damage and the maximum depth that had been obtained was 20mm. Time of Flight was measured on the sculpture zone [25]. The fused IRT and UV approach, which analyses and evaluates the condition of timber, was proposed by Kandemir et al. in 2007. The aforementioned procedure improved the alloy to make it suitable for long-term conservation programs. [12].

Rafael Fort et al. (2013) used a variety of techniques to evaluate two unique heritage structures. On the granite stone, a Schmidt hammer rebound experiment was conducted.

Table 1. Comparison of various methods and measurement

Methods	Parameter	Advantages	Disadvantages
Ultrasonic testing	Interfaces, voids, pores and fusion	Precise measurement	Control time is slow and the decision of explicit pores of every example
IRT	Surface temperature, discontinuities/interfaces, voids and debonding [30]	Control time is fast and a decent assessment of deformity measurement	Defect depth is not directly determined
Schmidt hammer rebound	Strength	A quick time of control constantly embraced every geometric kind	Slightly destructive technique
3D Terrestrial laser scanner	Voids, walls, columns and pillars in building structures, shapes and micro pores	Distance measured at every pointing direction, millimeter or sub-millimeter accuracies can be achieved	Care should be taken during the bundle adjustment
Digital Image Processing methods	Defective regions, shape, length and width of defective area	Fast analysis. Used in medical applications, video processing, robot vision	Require knowledge in software and the initial cost can be high depending upon the system used

The correlation coefficient between open porosity and water immersion were estimated as 0.99 [26].

Reis, H.C. et al. (2021) proposed the detection and classification of crack using deep learning techniques. The performance of proposed method is compared using SVM, Alexnet and decision tree. The accuracy of the proposed method that is ReCRNet provides better performance than existing methods [20]. Maria Auxiliadora et al. (2011) proposed a new method to detect the efflorescence of monuments in which the damage of a stone surface is classified and quantified. They also quantified the different weathering state of materials by efflorescence [27, 28].

Ramani P. et al. (2020) developed an image-processing method for the automatic detection of and classification of the crack decay. Crack is detected using combined canny with BHT and extracted the DWT features from the fused method. Crack length, location, width, orientation are measured using proposed method. Performance are evaluated in terms of accuracy, precision and selectivity, etc. This method improves the performance than existing methods [29].

3 Method

The image-based crack detection is described in Sect. 1. Accuracy-based detection is discussed in the second part.

3.1 Image Based Crack Detection

There are few challenges in image detection particularly due to poor lighting conditions and shading in the acquired images. The image based crack detection algorithms are simple and automatic in nature. Many algorithms had been developed for crack detection and they are canny edge detection, Bottom Hot Transform, Morphological approach and so on and so forth.

To distinguish exploration challenges and to accomplish better performances a point-by-point review must be done. It has been noted down that many researchers have been proposing many techniques that could identify cracks in decay monuments and road. Their various exploration works has been surveyed and analyzed. The association of the survey at first starts up with the image processing-based crack detection and other decay detection with non-destructive methods.

Crack and Other Decay Detection Using an Image Processing. This area defines the fundamental building structure of crack and other decay analysis and detection.

The schematic diagram of image processing for crack and other decay detection is shown in Fig. 1. The image of intrinsic sites from the high-resolution camera has been collected at first. Then, these images are preprocessed to remove the noise and resizing has been made. An improvement technique has increased the quality of the image, and crack and efflorescence decay features are extracted. Thus, the cracks are separated and geometrical characteristics of crack such as width, length, major axis, minor axis, perimeter and direction of propagation are measured. Thus in the above

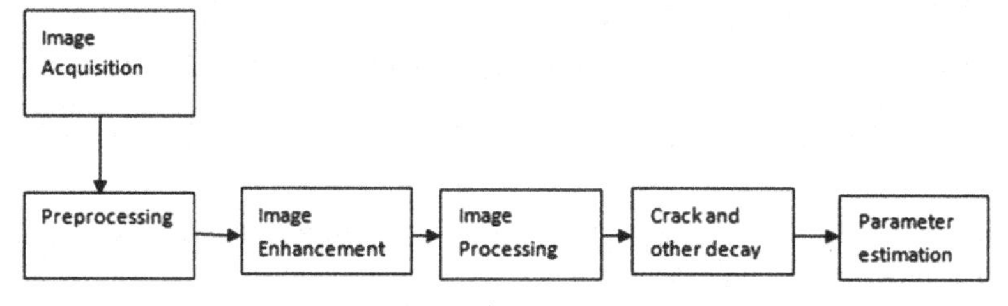

Fig. 1. Block Diagram of Crack analysis

schematic diagram the image processing approach for the detection of other decay and crack in monuments and other engineering structures has been described.

Heshan et al. (2018) suggested a new method for the estimation of crack depth using Make3D tool kit. The toolbox had been used to convert 2D images into 3D images. In this method, ground truth values were calculated from the laser scanner and used supervised learning to train and classify the model [31].

Abdel Qader et al. (2003) compared different crack detection methods using Fast Haar Transform, Canny filter and Sobel filter. The transform splits the image into a low-frequency component and a high-frequency component after identifying the edges of the surfaces with the frequency component [32]. Alam et al. (2015) used acoustic emission and correlation to locate cracks, with the latter technique also used to quantify internal damages. The K-means grouping is used for segmenting the image and analysis is done for three different beams [33].

The use of a Gabor filter approach by Salman et al. (2013) enhanced the effectiveness of crack detection. The crack is determined in all directions and 95% precision had been achieved in the proposed method [34].

Dare et al. (2006) developed a method to perform automatic crack detection in concrete structures. The bilinear interpolation method was used to process the crack pixel values and the sub pixels were measured using DOG filter [5]. Hweekwon et al. (2017) proposed a method to detect the cracks in the pressed board. Edge lines were removed and those outcomes are contrasted with the original sample board. With this technique the speed of crack detection and assessment had been improved a lot [35].

Tian Qinggue et al. (2019) developed a system that detects multiple cracks. In the proposed method the first step was combined edge detection and seed growing. The skeleton optimization procedure, which eliminated non-crack parts and returned the crack attributes, came next. The novel method retrieved multiple cracks from the concrete structure. This model distinguished the crack using a mixed input and classified the detected segments depend on the skeleton data. It was very useful in evaluating the solid structures [16].

The detection of crack particles was the subject of a model proposed by Lee et al. in 2009. By using nearest neighbour techniques, micro structural descriptors were identified, and crack features were obtained from the probability function. The edge effect was eliminated by the nearest neighbor [36]. Tang et al. (2016) created a wavelet-based multi-scale model, performed input image segmentation, and took clinical images for the examination. Edges were extracted utilizing WT and help to detail the immobile property. The technique had grouped all the pictures with Bayes hypothesis and accomplished the great effectiveness [37]. Parida et al. (2018) used a fuzzy clustering approach to extract the local variance features from an input image. The transition features were extracted from the OTSU Technique, and performed segmentation using hybrid method [38].

Aslam et al. (2020) presented a novel method for estimating the dimensions, measures, and shapes of metal flaws. Median filtering was used to eliminate the noise. With the use of CNN, segmentation and recognition of metal damage had been accomplished. A precision rate of 93% [39] had been achieved with the proposed work.

Sinha et al. (2006) introduced a statistical filter for examining the crack. Crack features were extracted with combined image. Cleaning and linking methods are used to segment the crack. They defeated their past work downside in the morphological activity [40]. Saini et al. (2017) provided a model to categorise and identify various monuments. DCNN was used to extract the presentation and the analysis was carried out with various precise perspectives and an exactness of 92% [41] had been achieved. The performance parameter of images for the crack detection which were discussed in the literature had been presented in the following section.

3.2 Accuracy Based Analysis

Thresholding is a basic method of segmentation and detection of the crack. The issue in thresholding is that both noise and crack has been detected as a crack. This is not an efficient method of detecting the crack. An accuracy of approximately less than 75% has been obtained through this. The numerous crack detection techniques are explained, and each technique is assessed by examining its related pros and demerits.

Addel-Qader et al. (2003) discussed the comparison of crack detection in bridge using Sobel, Fast Haar Transform, canny edge detection and Fast Fourier transform. In all the transforms filtering, compression and denoising had been done to remove the noise in the detected output. It had been concluded that the detection of crack using Fast Haar Transform provides more reliable output than other three methods [32].

Sheerin Sitara et al. (2018) used a preprocessing method to remove the noise in an image using wiener filtering. Singular Value Decomposition and Wavelet transform to overcome the non-uniform illumination of images had been proposed here. The length, area, number of cracks, and width of crack were calculated and thereby performing the automatic detection and classification of civil infrastructure techniques. However, using his method fine crack images could not be extracted [42].

Talab et al. (2016) described how to use image processing to find cracks in concrete constructions. The OTSU method had been employed to detect the major cracks and a Sobel filter had been used to remove the residual noise. Detection of minor cracks could not be achieved here [43].

Sankarasrinivasan et al. (2015) proposed combined HSV thresholding with Bottom Hat Transform for detection of a crack in civil structures. Integrated Unmanned Aerial Vehicles and Image Processing methods for crack detection have been utilized in the combined thresholding. This combined method determines both minor and major cracks. However, dimension of the structuring element and threshold value must be optimized in the entire process [44]. Salman, M. et al. (2015) introduced a novel method to automatically detect the crack in digital pavement images. The crack had been detected using Gabor filter in multidirectional dimension and thus allowed the detection of crack with 95% accuracy. However the computation time increases with the number of orientation [5].

Transform, morphological approach, Skeletonization procedures, and threshold methods are used to identify the majority of cracks. For identifying characteristics of the image, such as the length, width, and direction of the crack, the morphological technique is applied. The Combined method improves the accuracy of detected crack regions.

4 Results and Discussion

Transform, Morphological Approach, Skeletonization Techniques, and Threshold Method were used to detect the majority of the cracks. The shape of the image's features is determined using the morphological technique. The Crack length, width, and orientation are all measured. The discovered crack locations are more accurately identified using the Combined approach. The segmented output from the crack is shown in Fig. 2. Roberts, Sobel, and Prewitts and Morphological approach have greater noise. However, Canny edge detection and combination approaches produce less noise.

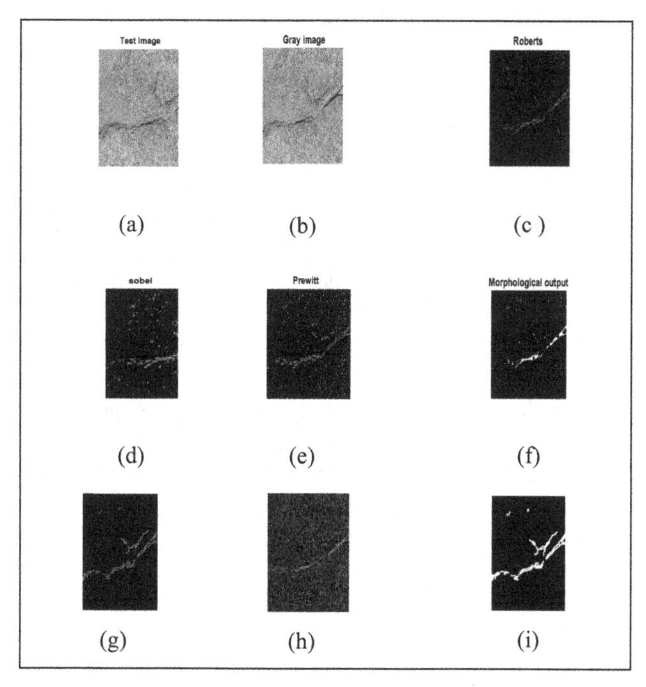

Fig. 2. (a) Input image, (b) Gray image, (c) Roberts method, (d) Sobel, (e) Prewitts, (f) Morphological output, (g) Canny edge detection, (h) BHT, (i) Combined bottom hot transform and canny edge detected

Table 2 compares several crack detection strategies with the suggested strategy employing image processing technologies. The different image segmentation algorithm for the detection of cracks such as Integrated Thresholding, Gabor filter, Enhanced Binarization, Optimized grey processing, Mixture of simple Gaussian density and minimum co-variance, Fast Haar Transform and Convolutional Neural Networks, Supervised approach with CNN was used in all existing method and which produces classification accuracy rate is less than 95% and U^2-Net based Deep Learning model performs pixel-level crack detection and produces 98% accuracy. The Combination of Integrated non destructive methods give better depth of crack than existing model.

Table 2. Comparison between various crack detection nondestructive methods

Existing	Nondestructive methods	Accuracy
Dhital D et al. (2012)	Ultrasonic scanning excitation and piezoelectric air coupled [45]	96%
Wang We et al. (2018)	Infrared thermal imager [46]	96%
Jun Yang et al. (2019)	Infrared thermal using CNN [47]	95%
Feng, Liuyang (2020)	Ultrasonic phased array [48]	Not specified
Aslam Y et al. (2020)	Supervised approach with CNN [40]	93%
Elhariri et al. (2022)	U^2-Net based deep learning [49]	98%

5 Conclusion

In this paper, the theoretical background is explained as per basic concepts from the literature review. In situ, Non-destructive testing (NDT) methods such as image processing methods are most suitable for weathering decay assessment. This method requires less time consumption, and it is user friendly. The depth of the crack cannot be estimated using IRT. The Larger size crack is difficult to analyze using Ultrasonic Testing. The damage severity was not estimated in existing methods. There is a lack of analysis of decay assessment and therefore the assessment of monuments needs more research to reach the ultimate goal of machine simulation of assessment. The deep Neural network method is to be used to structure and validate the results. The diagnostic process for determining the conservation status of building materials and monuments can be improved by the incorporation of non-destructive approaches.

Acknowledgments. We thank the anonymous referees for their useful suggestions.

References

1. Mohan, A., Poobal, S.: Crack detection using image processing: a critical review and analysis. Alex. Eng. J. 787–798 (2018). https://doi.org/10.1016/j.aej.2017.01.02

2. Fais, S., Casula, G., et al.: An innovative methodology for the non-destructive diagnosis of architectural elements of ancient historical buildings. Sci. Rep. 1–11 (2018)
3. Martinho, E., Dionísio, A., Almeida, F., et al.: Integrated geophysical approach for stone decay diagnosis in cultural heritage. Construct. Build. Mater. **52**, 345–352 (2014). https://doi.org/10.1016/j.conbuildmat.2013.11.047
4. Tavukcuoglu, A.Y.Ş.E., Caner-Saltik, E.N.: Mapping of visual decay forms and infrared imaging of stone structures for the maintenance and monitoring studies (1999)
5. Salman, M., Mathavan, S., Kamal, K., Rahman, M.: Pavement crack detection using the Gabor filter. In: 16th International IEEE Conference on Intelligent Transportation Systems (ITSC), pp. 2039–2044. IEEE (2013). https://doi.org/10.1109/ITSC.2013.6728529
6. Ghoshal, A., Ayers, J., et al.: Experimental investigations in embedded sensing of composite components in aerospace vehicles. Compos. B: Eng. **71**, 52–62 (2015). https://doi.org/10.1016/j.compositesb.2014.10.050
7. Montoto, M., Calleja, L., Perez, B., et al.: Evaluation in situ of the state of deterioration of monumental stones by non-destructive ultrasonic techniques. MRS Online Proc. Libr. Arch. **185**, 273–284 (1990)
8. Nappi, A., Cote, P.: Nondestructive test methods applicable to historic stone structures. Environ. Sci. Res. Rep. ES **20**, 151–166 (1997)
9. Tian, X.L., Zhou, X., Gao, F.: Nondestructive testing and analysis technology in the field of heritage preservation. Nondestruct. Test. 3 (2008)
10. Mishra, M., Bhatia, A.S., Maity, D.: Support vector machine for determining the compressive strength of brick-mortar masonry using NDT data fusion (case study: Kharagpur, India). SN Appl. Sci. **1**(6), 1–11 (2019). https://doi.org/10.1007/s42452-019-0590-5
11. Mishra, M., Bhatia, A.S., Maity, D.: A comparative study of regression, neural network and neuro-fuzzy inference system for determining the compressive strength of brick–mortar masonry by fusing nondestructive testing data. Eng. Comput. **37**(1), 77–91 (2019). https://doi.org/10.1007/s00366-019-00810-4
12. Pascale, G., Lolli, A.: Crack assessment in marble sculptures using ultrasonic measurements: laboratory tests and application on the statue of David by Michelangelo. J. Cult. Herit. **16**(6), 813–821 (2015). https://doi.org/10.1016/j.culher.2015.02.005
13. Jain, A.K.: Fundamentals of Digital Image Processing. Prentice-Hall, Englewood Cliffs, New Jersey (1989)
14. Schalkoff, R.J.: Digital Image Processing and Computer Vision, p. 286. Wiley, New York (1989)
15. Gheorghiu Bobaru, M., Păsăreanu, C.S., Giannakopoulou, D.: Automated assume-guarantee reasoning by abstraction refinement. In: Gupta, A., Malik, S. (eds.) CAV 2008. LNCS, vol. 5123, pp. 135–148. Springer, Heidelberg (2008). https://doi.org/10.1007/978-3-540-70545-1_14
16. Jung, H., Lee, C., Park, C.G.: Fast and non-invasive surface crack detection of press panels using image processing. Procedia Eng. **188**, 72–79 (2017). https://doi.org/10.1016/j.proeng.2017.04.459
17. Pragalath, H., Seshathiri, S., Rathod, H., et al.: Deterioration assessment of infrastructure using fuzzy logic and image processing algorithm. J. Perform. Constr. Facil. **32**(2), 04018009 (2018)
18. Costamagna, E., Quintero, M.S., Bianchini, N.: Advanced non-destructive techniques for the diagnosis of historic buildings: the Loka-Hteik-Pan temple in Bagan. J. Cult. Herit. **43**, 108–117 (2020). https://doi.org/10.1016/j.culher.2019.09.006
19. Fitzner, B.: Damage diagnosis on stone monuments-in situ investigation and laboratory Studies. Proc. Int. Symp. Conserv. Bangudae Petroglyph **7**, 29–71 (2002)

20. Fort, R., de Buergo, M.A., Perez-Monserrat, E.M.: Non-destructive testing for the assessment of granite decay in heritage structures compared to quarry stone. Int. J. Rock Mech. Min. Sci. **61**, 296–305 (2013). https://doi.org/10.1016/j.ijrmms.2012.12.048
21. Dong, J., Kim, B., Locquet, A., McKeon, P., Declercq, N., Citrin, D.S.: Non destructive evaluation of forced delamination in glass fiber-reinforced composites by terahertz and ultrasonic waves. Compos. B: Eng. **79**, 667–675 (2015). https://doi.org/10.1016/j.compositesb.2015.05.028
22. Cuccuru, F., Fais, S., Ligas, P.: Dynamic elastic characterization of carbonate rocks used as building materials in the historical city, centre of Cagliari (Italy). Geol. Soc. Lond. **47**(3), 296–305 (2014). https://doi.org/10.1144/qjegh2013-061
23. Brooks, A.J., Hussey, D.S., Yao, H., et al.: Neutron interferometry detection of early crack formation caused by bending fatigue in additively manufactured SS316 dogbones. Mater. Des. **140**, 420–430 (2018). https://doi.org/10.1016/j.matdes.2017.12.001
24. Garnier, M.L., Pastor, F., Eyma, L.B.: The detection of aeronautical defects in situ on composite structures using non destructive testing. Compos. Struct. **93**(5), 1328–1336 (2011). https://doi.org/10.1016/j.compstruct.2010.10.017
25. Fais, S., Ligas, P., Cuccuru, F.: Dynamic elastic characterization of carbonate rocks used as building materials. In: Second International Congress Dimension Stones (ICDS) XXI Century Challenge, pp. 319–323 (2008). https://hdl.handle.net/11584/105878
26. Kandemir-Yucel, A., Tavukcuoglu, A.Y.ŞE., Caner-Saltik, E.N.: In situ assessment of structural timber elements of a historic building by infrared thermography and ultrasonic velocity. Infrared Phys. Technol. **49**(3), 243–248 (2007). https://doi.org/10.1016/j.infrared.2006.06.012
27. Reis, H.C., Khoshelham, K.: ReCRNet: a deep residual network for crack detection in historical buildings. Arab. J. Geosci. **20**, 1–13 (2021)
28. Uchida, E., Ogawa, Y., Maeda, N., et al.: Deterioration of stone materials in the Angkor monuments, Cambodia. In: Developments in Geotechnical Engineering, vol. 84, pp. 329–340. Elsevier (2000).https://doi.org/10.1016/S0165-1250(00)80027-9
29. Vázquez, M.A., Galán, E., Guerrero, M.A., Ortiz, P.: Digital image processing of weathered stone caused by efflorescences: a tool for mapping and evaluation of stone decay. Construct. Build. Mater. **25**(4), 1603–1611 (2011). https://doi.org/10.1016/j.conbuildmat.2010.10.003
30. Ramani, P., Subbiah Bharathi, V.: Detection and classification of crack in heritage structures using machine learning techniques. J. Crit. Rev. **7**(19), 1595–1600 (2020). https://doi.org/10.31838/jcr.07.19.195
31. Jubouri, Q.A., Al-AzawiAl-Taee, R.J.: Efficient individual identification of zebrafish using Hue/Saturation/Value color model. Egypt. J. Aquat. Res. **44**(4), 271–277 (2018). https://doi.org/10.1016/j.ejar.2018.11.006
32. Shehata, H.M., Mohamed, Y.S., Abdellatif, M., et al.: Depth estimation of steel cracks using laser and image processing techniques. Alex. Eng. J. **57**(4), 2713–2718 (2018). https://doi.org/10.1016/j.aej.2017.10.006
33. Abdel-Qader, I., Abudayyeh, O., Kelly, M.E.: Analysis of edge-detection techniques for crack identification in bridges. J. Comput. Civ. Eng. **17**(4), 255–263 (2003). https://doi.org/10.1061/(ASCE)0887-3801(2003)17:4(255)
34. Alam, S.Y., Loukili, A., Grondin, F., Rozière, E.: Use of the digital image correlation and acoustic emission technique to study the effect of structural size on cracking of reinforced concrete. Eng. Fract. Mech. **143**, 17–31 (2015). https://doi.org/10.1016/j.engfracmech.2015.06.038
35. Dare, P., Hanley, H., Fraser, C.: An operational application of automatic feature extraction: the measurement of cracks in concrete structures. Photogram. Rec. **17**(99), 453–464 (2002). https://doi.org/10.1111/0031-868X.00198

36. Qingguo, T., Qijun, L., Ge, B., et al.: A methodology framework for retrieval of concrete surface crack′s image properties based on hybrid model. Optik **180**, 199–214 (2019). https://doi.org/10.1016/j.ijleo.2018.11.013
37. Lee, S.G., Mao, Y., Gokhale, A.M., et al.: Application of digital image processing for automatic detection and characterization of cracked constituent particles/inclusions in wrought aluminum alloys. Mater. Charact. **60**(9), 964–970 (2009). https://doi.org/10.1016/j.matchar.2009.03.014
38. Tang, W., Wang, Y., et al.: An image segmentation algorithm based on improved multiscale random field model in wavelet domain. J. Ambient Intell. Hum. Comput. **7**(2), 221–228 (2016). https://doi.org/10.1007/s12652-015-0318
39. Parida, P., Bhoi, N.: Fuzzy clustering based transition region extraction for Image segmentation. Eng. Sci. Technol. Int. J. **21**(4), 547–563 (2018). https://doi.org/10.1016/j.jestch.2018.05.012
40. Aslam, Y., Santhi, N., Ramasamy, N., Ramar, K.: Localization and segmentation of metal cracks using deep learning. J. Ambient. Intell. Hum. Comput. **12**(3), 4205–4213 (2020). https://doi.org/10.1007/s12652-020-01803-8
41. Sinha, S.K., Fieguth, P.W.: Automated detection of cracks in buried concrete pipe images. Autom. Constr. **15**(1), 58–72 (2006). https://doi.org/10.1016/j.autcon.2005.02.006
42. Saini, A., Gupta, T., Kumar, R., et al.: Image based Indian monument recognition using convoluted neural networks. In: International Conference on Big Data, IoT and Data Science (BID), pp. 138−142. IEEE (2017). https://doi.org/10.1109/BID.2017.8336587
43. Sheerin Sitara, N., Kavitha, S., Raghuraman, G.: Review and analysis of crack detection and classification techniques based on crack types. Int. J. Appl. Eng. **13**(8), 6056–6062 (2018)
44. Talab, A.M.A., Huang, Z., et al.: Detection crack in image using Otsu method and multiple filtering in image processing techniques. Optik Int. J. Light Electron. **127**(3), 1030–1033 (2016). https://doi.org/10.1016/j.ijleo.2015.09.147
45. Dhital, D., Lee, J.R.: A fully non-contact ultrasonic propagation imaging system for closed surface crack evaluation. Exp. Mech. **8**, 1111–1122 (2012). https://doi.org/10.1007/s11340-011-9567-z
46. Wang, W., Li, Q.: Crack identification of infrared thermal imaging steel sheet based on convolutional neural network. MATEC Web Conf. EDP Sci. **232**, 01053 (2018). https://doi.org/10.1051/matecconf/201823201053
47. Yang, J., Wang, W., Lin, G.: Infrared thermal imaging-based crack detection using deep learning. IEEE Access **7**, 182060–182077 (2019). https://doi.org/10.1109/ACCESS.2019.2958264
48. Feng, L., Qian, X.: Enhanced sizing for surface cracks in welded tubular joints using ultrasonic phased array and image processing. NDT E Int. **116**, 102334 (2020). https://doi.org/10.1016/j.ndteint.2020.102334
49. Elhariri, E., El-Bendary, N., Shereen, A.: Automated pixel-level deep crack segmentation on historical surfaced using U-net models. Algorithms **15**(8), 281 (2022). https://doi.org/10.3390/a15080281

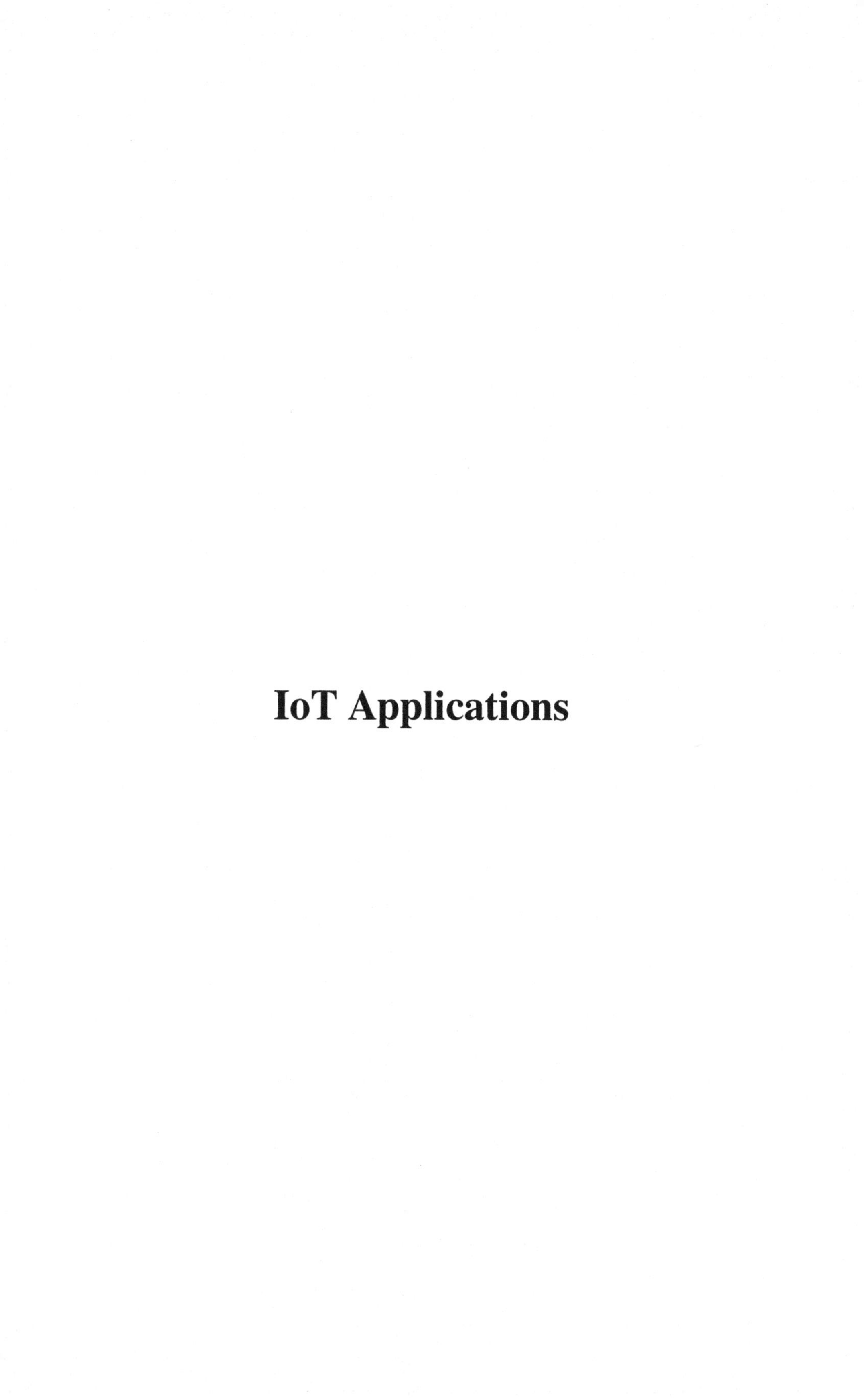

IoT Applications

IoT Enabled Driver Compatible Cost-Effective System for Drowsiness Detection with Optimized Response Time

Argha Sarkar[1], Mayuri Kundu[1], Prakash Pareek[2], Nishu Gupta[3](✉), and Manuel J. Cabral S. Reis[4]

[1] School of Computer Science and Engineering, REVA University, Bangalore 560064, Karnataka, India

[2] Department of Electronics and Communication Engineering, Vishnu Institute of Technology, Bhimavaram 534202, Andhra Pradesh, India

[3] Department of Electronic Systems, Faculty of Information Technology and Electrical Engineering, Norwegian University of Science and Technology, 2815 Gjøvik, Norway
nishu.gupta@ntnu.no

[4] Engineering Department, UTAD/IEETA, 5001-801 Vila Real, Portugal

Abstract. The present work researches driver drowsiness, which constitutes a huge problem, and can turn into fatal incidents potentially involving the losing of lives, being it while driving in a highway (car, bus, truck, etc.), in the railway, or any other transportation mean (ship, airplane, etc.). Recent technology has been involved in the study of the fatigue behaviour. The purpose of this work is to identify the in-driver's drowsiness, contributing to get rid of the accidents, mainly during the night, and to ensure better safety in train and on the highways. A camera is used to capture images from the driver, and face detection is executed in real-time to find out the exact position of driver's face. Here, the blinking of the driver's eye is under consideration. The fundamental concept lies in the closing of the driver's eye for a specific duration of time. If the closing time exceeds that certain duration, drowsiness is detected and an alarm is sounded for awareness. Python and OpenCV (using Haarcascade library) are the fundamental programming platforms for sensing the facial attributes. In this work, a cost-effective system for the driver's drowsiness detection is aimed. Further, the result also indicates the reduction of response and processing time, which is ideal for real time detection.

Keywords: IoT · Drowsiness · Detection · OpenCV

1 Introduction

The importance of vehicle safety has increased with the advancement of technology in all spheres of life, and the main focus is now on reducing the alarming time when an accident occurs so that the rescue team can treat the injured faster. When we try to go deep into the root causes of the road fatalities, drowsiness of the driver emerges as one of

N. Gupta et al. (Eds.): IC4S 2022, LNICST 472, pp. 159–169, 2023.
https://doi.org/10.1007/978-3-031-28975-0_13

the prime causes. This drowsiness is caused by the intake of unwanted drugs or mental fatigue of driver. The number of accidents is expected to decrease drastically if the eye movement of the driver can be monitored. It involves the recognition of human face efficiently in a timely manner which is a complex problem.

The human face is lively and active with a high grade of changeability. In computer vision, it is a very difficult problem to perform face recognition in real time dynamic environments. Human eyes play a major role during face recognition as well as facial expression analysis. Salient features dynamically change while a person feels drowsiness. Eyes are the most relevant and fairly stable landmark on the face in contrast with other facial attributes. Therefore, it is more significant to detect eyes before recognizing other facial attributes. The eye position can estimate the location of other facial elements/features. Besides, the shape, size, position and the rotation of the image-plane of the face can be normalized by the location of both eyes [1–5].

The basic contributions of the present work are directed to:

- Be able to accurately detect a face from an image.
- Be able to detect the region of interest, in this case the eyes.
- Accurately classify the status of the eye either closed/ open.
- Provide a warning to the driver if drowsiness is detected.
- Provide cheaper systems that are affordable to the low-income people.

The rest of the paper is organized as follows. Section 2 provides stepwise methodology adopted in this work. It is followed by a brief explanation of implementation which is depicted in the section. Finally, the attained results are analysed and shown in Sect. 4. The conclusion of this work is provided in last section.

2 Methodology

As stated before, the main objective of the present work is to detect the drowsiness of a driver (driving a car, train, etc.). It can be estimated by the duration of the eye closing time [5, 6]. If the eyes are in a closed state for a certain period of time, an awareness notification sound will be ringed. In this work, faster response and detection, irrespective of specific drivers, are the fundamental objectives. The number of frames associated with eye closure are monitored. If the number of frames reach a specific value, then an awareness message is displayed on the screen showing the notification "driver is feeling drowsy".

The following algorithm is implemented. Initially, the image is captured by a webcam and sent for further processing. The Haarcascade file face.xml is used to find and sense the faces in an individual frame. If the situation arises that no face is detected, then the next frame is acquired. On other hand, if a face is recognized, then a region of interest (RoI) is declared within the face. This RoI contains both eyes. The purpose of declaring a RoI is to reduce the computational effort of the system. Finally, by using Haarcascade_eye.xml, the eyes are detected from the RoI. Figure 1 depicts the overall representation of working of the proposed system.

In the next subsections a brief explanation of the main steps of the algorithm, presented in Fig. 1, is presented.

2.1 Image Acquisition

```
# Initiation of the video stream thread
print("[INFO] starting video stream thread...")
vs = VideoStream(src=0).start()
# vs = VideoStream(usePiCamera=True).start()
time.sleep(1.0)
```

The function “vs = VideoStream(src=0).start()” grabs the video streaming from a USB camera. There is a second sleep to warm up camera sensor.

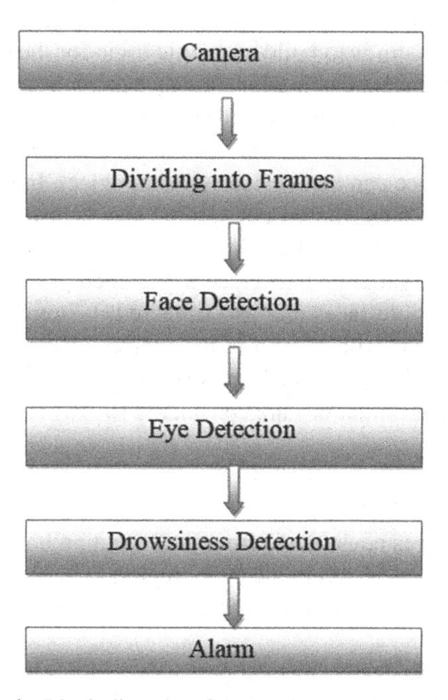

Fig. 1. Block diagram of the implemented algorithm

2.2 Dividing into Frames

Step 1. while condition is True:
frames from video stream are collected
Step 2. convert the collected frames into grayscale channels after resizing
Step 3: face detection takes place in the converted frames

In practice, in a real time scenario, the system deals with captured video which is in turn processed further. But here, the mentioned algorithm is compatible for image only. Therefore, the captured video is segmented into frames for the corresponding analysis.

2.3 Face Detection

In this specific stage, the aim is to build a specific logic for the recognition of the face. By face recognition, it is meant that marking the face in a frame, i.e., finding the position of facial attributes. The frame is supposed to be random in nature. The detection is very selective to the facial related features and does not include other structures or objects in the surroundings. A face can be considered as an object. In such cases, the location of the target object is intended to find out and the size is also approximated to known particular class. The face detection algorithms are mostly based on recognizing the front side of the face. Herein, a situation like a face in the tilted position, or positioned in any other way, and also multiple faces are also triggered. It implies that it may solve the purpose of the rotation axis with respect to the present observer from the reference of the face in a particular position, or even if there is a vertical rotation plane. In the present algorithm, all the real time situations are considered [6–9].

Step 1: Repeat for (x,y,w,h) in rects:
face in detection in converted frames.
Step 2: a rectangle object is constructed from the bounding box of Haarcascade
Step 3: facial landmark is detected in the facial region with (x,y) co-ordinate
Step 4: facial landmark (x,y) co-ordinates is converted to array

The above function basically draws a rectangle in the image to point out the corresponding corner points. The rest of the parameters are to illustrate the colour and thickness and type of lines in the rectangle and visualize the coordinates.

The Facial Landmarks Predictor

Step 1. Initialize dlib's face detector
Step 2. facial landmark predictor is created with a message "("[INFO] loading facial landmark predictor…")"
Step 3. Indexes for left and right eye is grabbed from facial landmarks

2.4 Computing the Euclidean Distances and Expression for Ear He Facial Landmarks Predictor

Step 1. Two sets of Euclidean distance between left and right eye is computed from vertical eye landmarks (x,y) – co-ordinates and store it to A and B

Step 2. Euclidean distance between left and right eye is computed from horizontal eye landmarks (x,y) – co-ordinates and store it to C

Step 3. Eye aspect ratio is calculated using A ,B and C value and store it to ear(Eye Aspect Ratio)

```
ear= (A+B)/(2.0*C)
```

Step 4.
```
if(ear>x):
        x=ear
else if (ear<y):
        y=ear
return ear
```

2.5 Set Unique Threshold Value

EYE_AR_THRESH = x + ((y − x)/5)
EYE_AR_CONSEC_FRAMES = 48

2.6 Set Image Region of Interest

It extracts the left and right eye coordinates and compute the EAR (Eye Aspect Ratio), which is the average of Eye Aspect Ratios of left and right eyes.

Step 1. Extract left eye co ordinates
Step 2. Extract right eye co ordinates
Step 3. Left Eye Aspect Ratio (leftEAR) is computed from Left eye co-ordinates
Step 4. Right Eye Aspect Ratio (rightEAR) is computed from Right eye co-ordinates
Step 5. Calculate the average eye aspect ratio (ear)

ear = (leftEA + rightEAR)/2.0

2.7 Visualization of Eye Regions on a Frame et Image Region of Interest

Step 1. The convex hull is computed for each eye
Step 2. Visualize each of the eyes to plot contours for both the eyes.

2.8 Drowsiness Alert Based on Eye Aspect Ratio

Here, the ear is compared with the EYE_AR_THRESH – if it goes below the threshold, eyes are closed. COUNTER is incremented and subsequently verified to know the eye closure time for enough consecutive frames, to blow the awareness alarm. If the alarm is in off state, it will be turned on to notify the drowsy driver.

```
 if eye aspect ratio (ear)<blink threshold
                  then increment the blink frame counter by 1.
                  if eyes are closed in maximum number of frames
                           then alarm= ON
if args["alarm"] > 0:
         th.buzzer.blink(0.1, 0.1, 10,background=True)
draw an alarm on the frame

else:
                  reset the counter as 0 and alarm
```

The dip in eye aspect ratio indicates driver's drowsiness.

3 Implementation

As stated above, to implement the principles and algorithm presented in the previous section, we have used Python and OpenCV (using Haarcascade library).

Here, the purpose of using OpenCV is mainly to detect the image. OpenCV [OpenCV] is an open-source computer vision library which is shown in schematic form in Fig. 2. Additionally, OpenCV is chosen basically because of its computational efficiency in real-time image recognition, and "Simple-to-use computer infra" is one key feature of OpenCV which results in the design of highly sophisticated and fast response vision applications.

In a more practical implementation perspective, the reasons behind choosing OpenCV are:

- Specificity: In OpenCV, all the functions and data structures are designed aiming to the image detection.
- Efficiency: One of the alternatives is MATLAB, but consumes huge system resources, whereas, OpenCV, requires only 10MBytes RAM for a real-time application.
- Speediness: By using MATLAB, a maximum of 4 to 5 frames can be processed in every second. In contrast, OpenCV can process around 30 frames per second. It makes OpenCV more applicable to real time situations.

Visualization of facial landmark coordinates is crucial here, and thus eye aspect ratio (EAR) is a priority for drowsiness detection to obtain the desired accuracy. The critical parameters of eye status are classified based on the EAR. The level of opening and its landmark positions are extracted from an image: (a) extracting exact landmark position of eyes (b) calculating distance between vertical horizontal coordinates.

From the given Fig. 3, A is the distance between 1 & 5. B is the distance between 2 & 4 and C is the distance between 0 & 3. Relating A, B, and C distances by an appropriate formula. The above process of calculation is performed for both eyes after getting the image from video or target face. Then the eyes are detected and the corresponding landmark of both eyes.

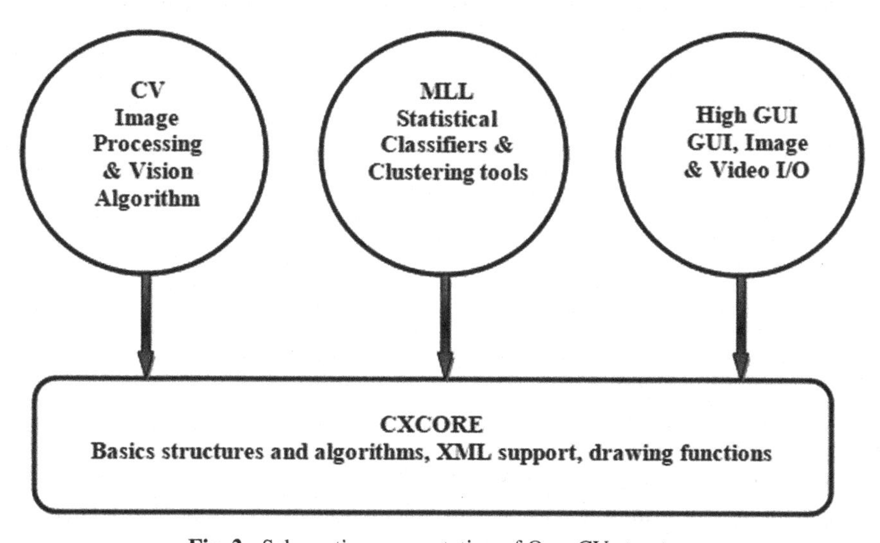

Fig. 2. Schematic representation of OpenCV structure

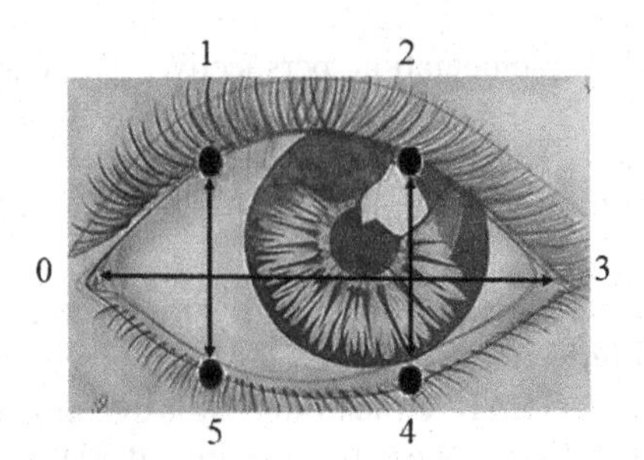

Fig. 3. Extraction and calculation of the exact landmark position of eyes

If this EAR value goes low compared to a unique threshold value, which automatically gets adjusted for every individual, an alarm is triggered. The novelty of the work lies on the automatic adjustment of the threshold value for every individual. One of the most important difficulties that we have had to ensure the effective drowsiness detection involving eye aspect ratio was the fixed threshold value, implying that the system was static, i.e., treating everybody equally. However, every individual can have unique threshold values based on their eye-opening area, and hence we have adopted a dynamic behaviour of our system, adapting the threshold value. This is one of the most important advancements in this work. Irrespective of a specific driver, the system calculates the threshold value for each and every driver dynamically.

The software-hardware interactions involved for each stage of the methodology are as follows:

Software components:

- Raspbian Stretch
- Python3
- Cmake Dlib
- OpenCV

Software implementation:

- Eye detection
- Detecting the eye center
- Determining pixel intensity of center
- Eye state determination
- Driver drowsiness detection with OpenCV
- Programming performed in python language and OpenCV using the Haarcascade library
- Drowsiness analysis

Hardware components:

- Raspberry Pi 3
- Logitech c270
- Buzzer

Hardware implementation:

- Implementation using LAPTOP and WEBCAM.
- Implementation using Raspberry Pi and Webcam.

Various steps and its related integration to the final drowsiness detection method and device are explained below:

Step 1: Camera grabbing: video streaming
Step 2: Dividing into frames: grabbing of the frame resized and convert ed it to grayscale.
Step 3: Face detection: sent for face detection and a pause is initiated for a second and then loop over frame in video stream.
Step 4: Facial detection: detection of facial landmarks, conversion of facial landmark (x, y) coordinates to the numpy array.
Step 5: Eye Detection: extraction of left and right eye coordinates, and estimation of threshold value.
Step 6: Eye aspect ratio (EAR) calculation: The coordinates are used to compute eye aspect ratio.
Step 7: Comparison: EAR with unique threshold value.
Step 8: Triggering of Alarm: if EAR < threshold value (signal is sent from raspberry pi to speaker)

4 Result and Analysis

To confirm the accuracy of the system to determine the eye blinks and drowsiness, a huge set of samples are considered from live video.

In this work, a USB webcam of 5 megapixels is connected to Raspberry Pi 3 model B+. A piezoelectric buzzer is incorporated in the system to produce the notification sound to make the driver conscious. It happens only when the duration of the closing of eyes reaches a certain threshold. The whole setup is tested in different real time situations for different drivers. High accuracy is achieved for the driver who is wearing spectacles also. The different samples with different conditions are shown in Fig. 4.

This system is appropriate to use in vehicles like bus, taxi, train, etc., to avoid accidents due to the driver's drowsiness. Sample outputs are shown below where such conditions are considered.

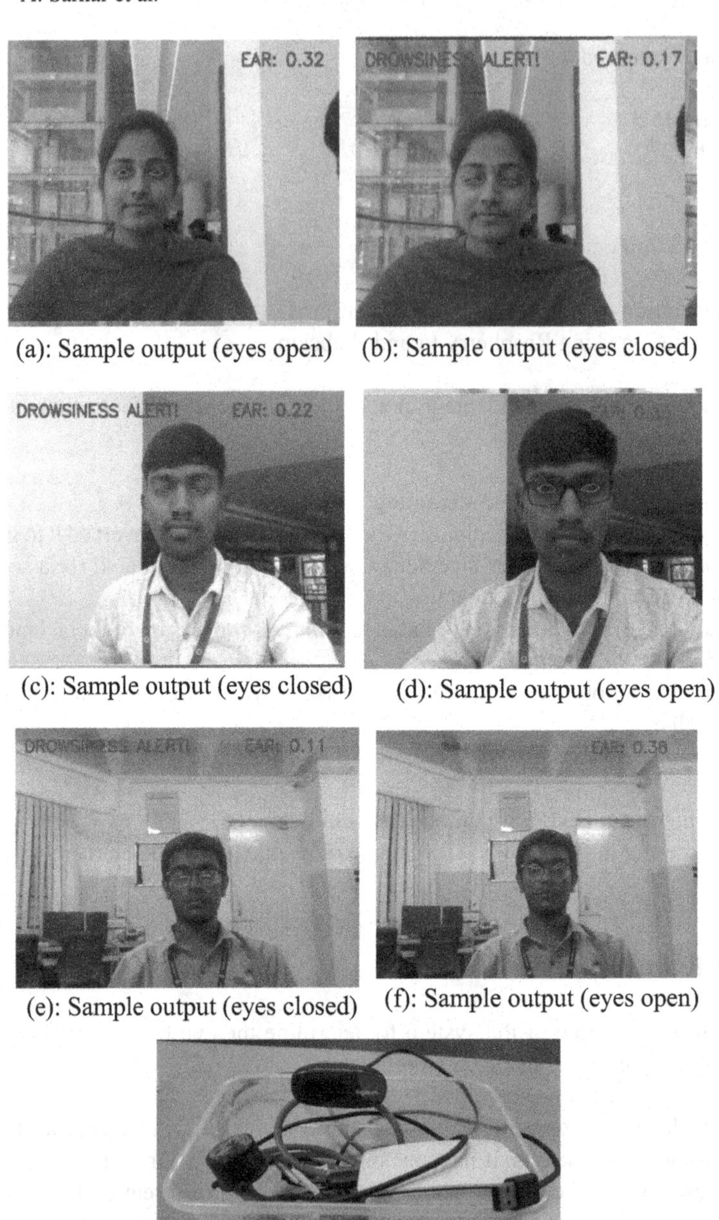

(a): Sample output (eyes open) (b): Sample output (eyes closed)

(c): Sample output (eyes closed) (d): Sample output (eyes open)

(e): Sample output (eyes closed) (f): Sample output (eyes open)

(g): Drowsiness detection-Device prototype.

Fig. 4. (a): Sample output (eyes open). (b): Sample output (eyes closed). (c): Sample output (eyes closed). (d): Sample output (eyes open). (e): Sample output (eyes closed). (f): Sample output (eyes open). (g): Drowsiness detection – device prototype.

5 Conclusion

We have presented a simple methodology to detect drowsiness in a driver. We have also presented a system to implement this methodology. This system includes a process able to detect the region of interest, in this case the eyes aspect ratio. A unique technique is used to classify the state of the eye, either closed or open, based on the programming performed using the Python language and OpenCV using the Haarcascade library. The proposed driver drowsiness detection system automatically adapts to the different drivers' characteristics, and is not limited to a specific driver. The system also provides high accuracy for the driver who is wearing spectacles. The system has minimized response and processing time, which is highly desirable for real time detection, and it can be considered a low-cost of cost-effective system, affordable to the low-income people.

Acknowledgments. The author would like to acknowledge Sree Vidyanikethan Engineering College Tirupati, India for providing the infrastructural support to pursue the work.

References

1. Sanjay, S., et al.: Drowsiness detection with OpenCV. In: Proceedings of 2021 Second International Conference on Electronics and Sustainable Communication Systems (ICESC), pp. 1421–1425. IEEE, Coimbatore (2021)
2. Khunpisuth, O., et al.: Driver drowsiness detection using eye-closeness detection. In: Proceedings of 2016 12th International Conference on Signal-Image Technology and Internet-Based Systems (SITIS), pp. 661–668. IEEE, Naples (2016)
3. Thulasimani, L., Prithashasni, S.P.: Real time driver drowsiness detection using opencv and facial landmarks. Int. J. of Aquat. Sci. **12**(2), 4297–4314 (2021)
4. Chirra, V.R.R., Uyyala, S.R., Kolli, V.K.K.: Deep CNN: a machine learning approach for driver drowsiness detection based on eye state. Rev. d'Intelligence Artif. **33**(6), 461–466 (2019)
5. Satish, K., et al.: Driver drowsiness detection. In: Proceedings of 2020 International Conference on Communication and Signal Processing (ICCSP), pp. 380–384. IEEE, Chennai (2020)
6. Cech, J., Tereza, S.: Real-time eye blink detection using facial landmarks. In: Proceedings of 21st Computer Vision Winter Workshop. Rimske Toplice (2016)
7. Mohanty, A., Bilgaiyan, S.: Drowsiness detection system using KNN and OpenCV. In: Swain, D., Pattnaik, P.K., Athawale, T. (eds.) Machine Learning and Information Processing. AISC, vol. 1311, pp. 383–390. Springer, Singapore (2021). https://doi.org/10.1007/978-981-33-4859-2_38
8. Bergasa, L.M., et al.: Real-time system for monitoring driver vigilance. IEEE Trans. Intell. Transport. Syst. **7**(1), 63–77 (2006)
9. Cech, J., Franc, V., Matas, J.: A 3D approach to facial landmarks: Detection, refinement, and tracking. In: Proceedings of 2014 22nd International Conference on Pattern Recognition, pp. 2173–2178. IEEE, Stockholm (2014)

Voice Based Objects Detection for Visually Challenged Using Active RFID Technology

K. V. S. H. Gayatri Sarman and Srilakshmi Gubbala(✉)

Vishnu Institute of Technology, Bhimavaram 534202, AP, India
{gayatrisarman.k,Srilakshmi.g}@vishnu.edu.in

Abstract. Now a days people are busy with their daily activities so that they can be forgotten about the regular usable items like bike or car keys, eye glass, books and bag are required for our needs but usually they are misplaced the objects and forgotten either in a home or office. For example shelves, sofa gaps and table desks etc., and also visually impaired people are having trouble for finding these objects. This trend has led to an increasing demand for devices and services. Perception of remote objects unravels the above mentioned problem That assists visually impaired persons in living comfortably and independently, as well as normal people in recognizing objects. In this paper, the proposed electronic gadget is based on voice recognition method. It consists of transmitter and receiver chip, voice recording and playback chip and VR module and RF Tags. Utilization of this gadget is very easy because this is operated through voice commands instead of pressing buttons.

Keywords: VR module · Transmitter and Receiver chip · Recording playback chip · Arduino · Easy VR Commander

1 Introduction

Interaction between visually impaired people with the surrounding objects at different places is very difficult. It is required to identifying the objects simple and intuitive. In this paper, we proposed a sophisticated method to meet the goal. Identification of objects is done by using radio frequency technology and it is extensively used in the many industries due to its less economic [1]. This RFID device assist the visually impaired people as well as old people to recognize the misplaced objects in their home or office [2, 3] and Using this RFID Technology we can recognize object's locations in a reliable, flexible, inexpensive and scalable manner [4]. It is difficult to recognize different types of objects at different locations at the same time due to collision. Nowadays many Technologies are available to identifying the objects i.e., Bar code, RFID etc., but they have their own limitations [5]. For the past two decades RFID Technology has been popular to identifying the remote objects [6]. RFID Technology have three categories 1. Passive RFID (Radio Frequency Identification) 2. Active RFID (Radio Frequency Identification) 3. Semi active RFID (Radio Frequency Identification). Due to long range applications we

N. Gupta et al. (Eds.): IC4S 2022, LNICST 472, pp. 170–179, 2023.
https://doi.org/10.1007/978-3-031-28975-0_14

select an active RFID Technology in our proposal. The theme of this paper is to Perception of Retractus Objects through Voice based Active RFID Technology. In this proposal we divide entire system into two categories i.e., RFID Reader with voice module and RFID tags and user interface. Each and every object contains unique coded Tag. RFID reader will take care about the tracking of Tags information. These tags are scan by RFID Reader at a rate of hundreds per second [7]. Here the RFID Reader range is typically 100 m. This Technology is most suitable for Indoor navigation [8].

2 Literature Survey

Now a day's RFID technology is used for various application in three different ways, Those are Passive RFID, Active RFID and Semi Active RFID. Among all these Active RFID is used for long distance communication [9].

In 2009, RFID for Identification of stolen/lost [10] items proposed that the private data regarding the manufacturing details stored in RFID tag can be test a security framework designed to authenticate RFID readers before allowing them to access.

In 2014, The misplaced objects are identified using active RFID. It reduces the user time to detect the misplaced object [11], Such as medicine box, mobile, key chain etc. with in the room vicinity. Even though RFID technology is a old, but nowadays it is used in many applications. In this Faraday's laws of electromagnetic principles are used ti detect the misplaced objects Maxwell's equations describing the electromagnetism. These principles are foundation for modern RF communication.

In 2021, RFID based attendance system is proposed which is useful for colleges, schools and corporate offices etc. The manual attendance system is difficult to manage and time consuming. This RFID based attendance system overcome these drawbacks in efficient manner [12].

Using an object finder, one can identify lost household and personal items in a home [13]. A locator, like those offered by specialist retailers, has an interrogator with a few buttons of different colours and a tag of the same colour as each button. By adding a tag to an item that needs to be tracked, the user can find the item by pressing the button on the interrogator that corresponds to the tag. The item's tag makes the appropriate noises and flashes, enabling the user to locate the object.

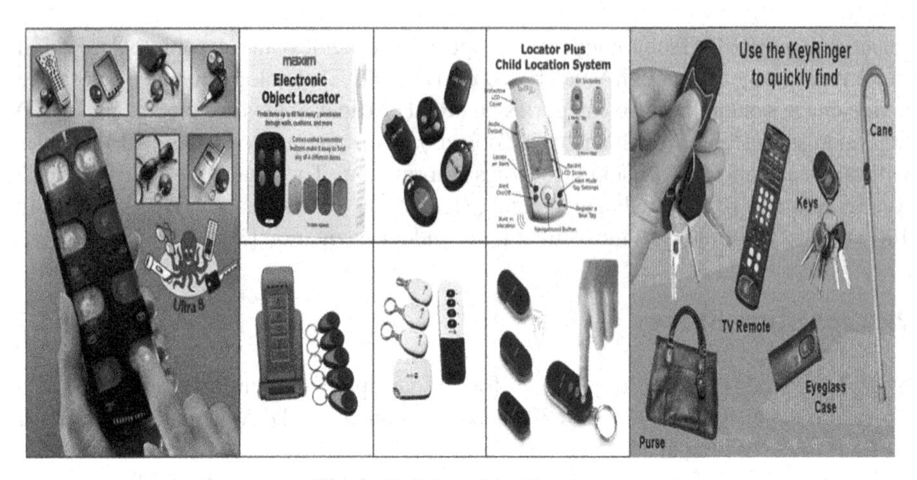

Fig. 1. Existing object locators

Here Fig. 1 shows existing RFID-based object locator [14]. These object locators are extensible, reusable and low support. They are simple for users to set up and use. This examination shows that quest investment utilization for all plans and look through plans depend the abilities of RFID readers and RF transceivers used by agents.

3 Propose Methodology

This paper gives a voice based object finder framework for visually challenged persons in home applications RFID Reader have Arduino Microcontroller, RF Transmitter along with Voice recognition (VR) module as shown in Fig. 2. RFID Tag comprises of PIC Microcontroller, Buzzer and RF Receiver as shown in Fig. 3. Whenever Voice Command is given by the user VR module in the RFID Reader recognizes and send corresponding digital code to the Microcontroller. The digital code coming from the VR Module is pre-recorded in the memory of the VR module. By using Time Division Multiplexing Process, Micro controller will send unique ID Code to all the Tags at a particular time interval. By this process all the Tag ID codes are sending in different time intervals. This Technique will avoid the collision of the data coming from the RFID Reader. At the same time Microcontroller activates the Voice Module which announces voice for the sake of cross checking of the correct voice Command. All the RFID Tags receives the Unique ID Code transmitted by the RFID Reader at a particular time interval. But only matched Tag gives response and sounds the buzzer until the user picks up and RESET.

The voice-based navigation system block diagram for individuals who are blind is shown in Fig. 2. RFID reader is connected to Arduino UNO. The Tag receives voice commands from blind people. At each location, passive RFID tags are positioned. Through the use of Arduino, the voice command [15] reaches its goal. The RFID reader reads the data from the RFID Tag when the RFID label enters its field of vision. The device is turned on after the antenna receives power from the RF reader. The information contained in the chip is then recovered and sent back to the reader at that point. Assuming the RFID tag and reader are in sync, this shows that the goal has been reached in the.

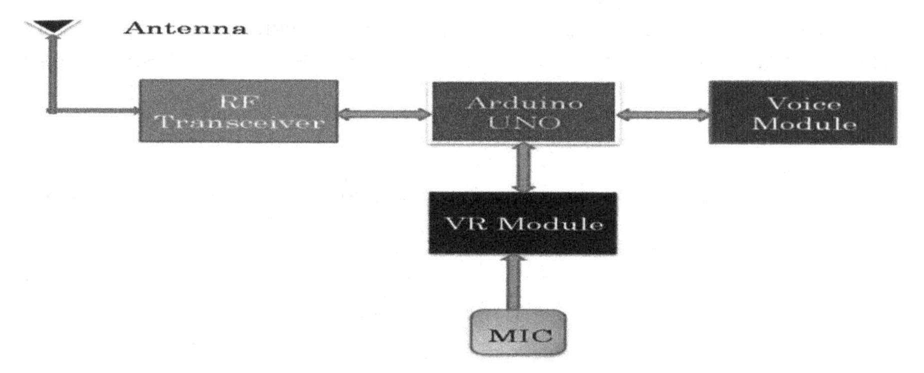

Fig. 2. Block Diagram of RFID Reader

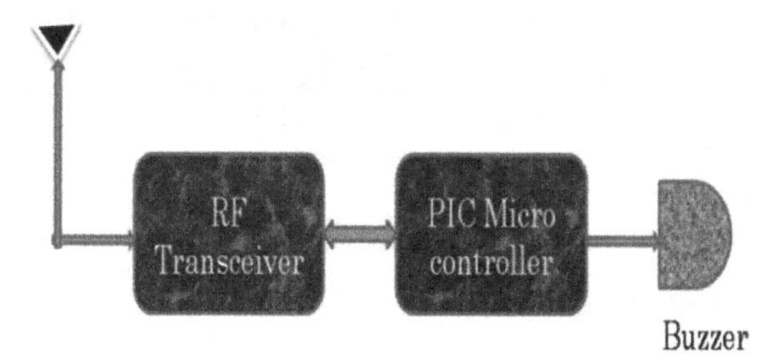

Fig. 3. Block Diagram of RFID Tag

In existing methods they are used buttons to recognize the remote object but in our proposed methodology we used voice commands to recognize the objects and Micro controller decodes the voice commands and activates the voice chip that announces the voice command for verifying the user command as shown in Fig. 4.

At the same time micro controller sending the corresponding unique code to all the tags in a time division multiplexing process. These RFID Tags are connected to the different objects for identification purpose. Each RFID Tag consisting of RF Receiver, Micro controller. It has a distinct reader ID that distinguishes it from the rest. Here there are 5 RFID Tags for identifying Five different objects. It receives Transmitted ID from the RFID Reader and compares with its own unique ID, If both ID's are matched then it will sound the buzzer until RESET by the user. The Tags are kept around 50 m diameter from the reader. This RFID Tags are working with a battery-powered power source As a result, the range of these sensors is greater than that of passive sensors. It is possible to increase the number of sensors so that we can identify more objects.

The voice module consisting of PIC16F877A and APR9600 chips. The interfacing between PIC 16F877A and APR9600 is through parallel interface. In our proposed method we have programmed the PIC 16F877A to access the voice chip (APR9600) through the corresponding addresses.

We have entered these commands through PC hyper terminal through RS232 interface with a Baud rate of 9600 voice signals are pre recorded in the APR9600 through Mic in the Recording mode.

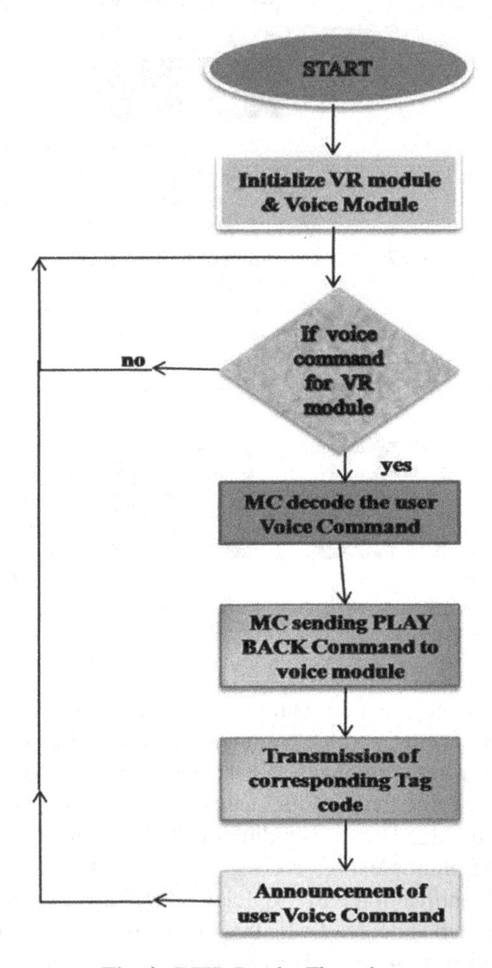

Fig. 4. RFID Reader Flow chart

Voice module interfaced with other microcontrollers through RS232 Interface. Voice module is an integration of PIC16F877A Microcontroller and APR9600 Voice recording and play back chips. Here PIC16F877A Microcontroller is programmed for a dedicated accessing of APR9600 as shown in Fig. 5.

Main theme of This Microcontroller is to form bridge between APR9600 with outside Microcontroller. In this we have used 434 MHz RF Transmitter and Receiver modules for the wireless communication between the Reader and Tags. We have used parallel interfacing between the APR9600 and PIC16F877A. We can prerecord the voice signals in a particular location in APR9600 by grounding the corresponding address line through

PIC Microcontroller during recording time. Playback can be performed by grounding the particular address line at once. APR9600 contains 8 address lines are used for recording and play back purpose.

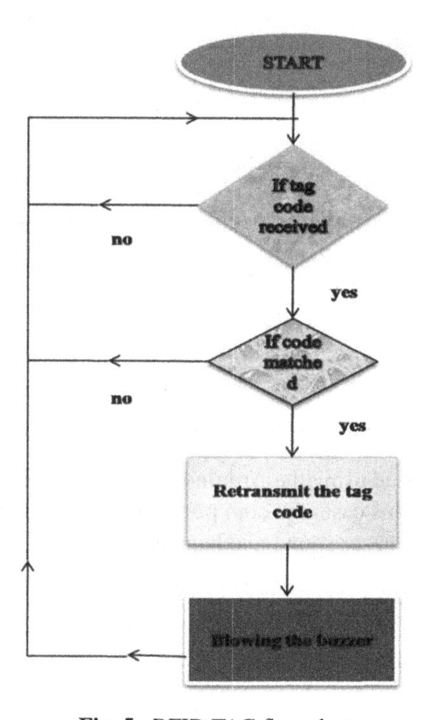

Fig. 5. RFID TAG flow chart

The positioning accuracy of the active RFID Tags is in the range of 10 m. The maximum signal range and accuracy can be increased by using trilateration. ITU (International Telecommunication Union) model is useful for indoor environments. Path loss within a room or a confined space within a building might be bounded by any type of wall. The measured RSSI (Received signal strength indication) and the range from the transmitter have a logarithmic relationship. The can be expressed mathematically as shown in Eq. (1)

$$x_T = 20log_{10}f_c + 10.m.log_{10}D - 28 \tag{1}$$

where x_T is the total signal strength in [dBm]

f_c is the carrier frequency in [MHz].

m is the signal strength exponent.

D is the range between the RFID tag and the RFID reader in [m].

A new, simplified equation that uses three fixed parameters from Eq. (2) can be used to determine the ideal parameters for an RFID system.

$$x_T = b_0 + b_1.log_{10}D \tag{2}$$

where b_0 and b_1 are coefficients found during calibration using measurements on a known baseline.

Then the parameter Eq. (3)

$$b_0 = 20log_{10}f_c - 28 \tag{3}$$

Equation (4) may then be used to calculate the distance d between the RFID tag and the RFID reader: with the coefficients

$$D = 10^{\frac{x_T - b_0}{b_1}} \tag{4}$$

3.1 Testing Method

This proposed work is tested using the Arduino communication interface. When a voice command is provided in this case, the com port will display the information. The associated object will emit a buzzer sound and have its LED light up in response to a voice command.

4 Implementation and Results

Fig. 6. Interfacing voice recognizer with Arduino UNO

In the above Fig. 6 We have Interfaced Easy VR shield with ARDUINO UNO for the voice commands recognition.

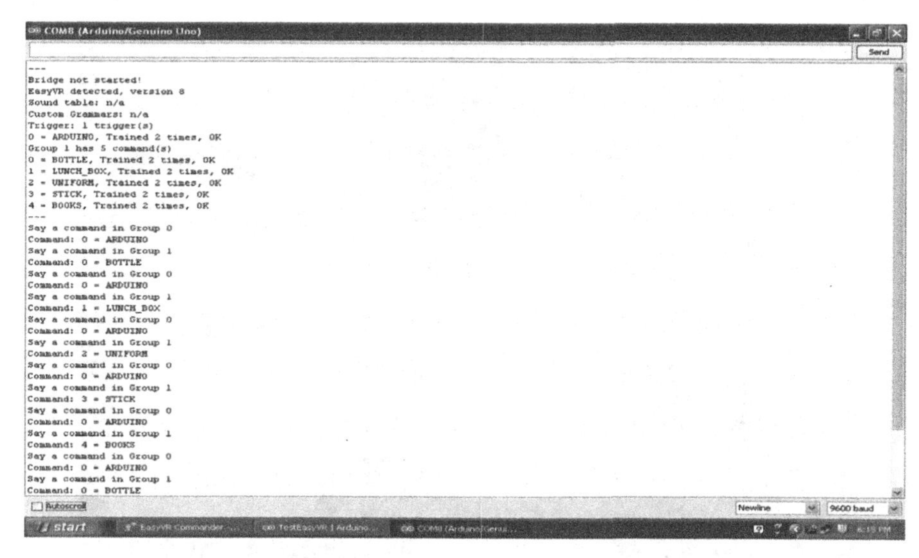

Fig. 7. Easy VR module simulation results

The Fig. 7 shows the simulation results of Easy VR modules in training mode and recognition mode. In the Group 0 of Easy VR module we have stored customized voice commands. The commands location information is as follows.

0 - location **"Bottle", 1 -** location **"Lunch Box",** 2 - location **"Uniform",** 3 - location **"Stick",** 4 - location **"Books"**

In the bottom portion of the above Fig. 8 shows recognition of the voice commands. Whenever voice command matches with the data base information it sends in the ASCII text format. As shown in Fig. 6, five different objects are identified by the visually challenged people by using our prototype model with in the 50 m range. Here we have attached 5 different RFID Tags with unique ID. This unique ID code is written in one array in ASCII format of that Tag Microcontroller. This written ID code is compared with Live receiving ID code from the Reader, when both are matched the buzzer will ringing sound. In Easy VR Module we recorded predefined voice commands using Easy VR commander software through PC.

Fig. 8. RFID Tags attached to the different objects (Bottle, Lunch Box, Uniform, Stick, Book)

5 Conclusion and Future Scope

Our Proposed methodology shows reliable, accurate, streamlined and scalable RFID-based approach for visually challenged people to identify their objects with in the room. In present existing system they have button type technique. To improve convenience of the user especially for visually challenged people we incorporate voice recognition methodology. So that user can easily operate device. Active RFID Tags track the objects over long distance comparing with the Passive RFID Tags. Time division Multiplexing approach leads to add more number of RFID tags to our system. We can extend this work for identifying objects in 3D regions, multiple stationary, outdoor environments mobile object localization.few limitations of RFID is the regular maintenance of Batteries, signal will lose due to any metal obstacles. Future research of this work can be additional improvements in overall speed, particularly for moving objects.

This paper points on enhancing a visually impaired acknowledgment framework through utilizing RFID and covers a concise understanding of important work about the RFID framework. The audit of RFID framework presents the specialized elements of this utility and presents the idea of exploration being created invoice based for blind individuals. The data, for example, the expenses of the parts of the proposed RFID-based framework is given. RFID innovation is maturing and extending as the world dynamically moves towards a contact-less climate. All things considered, the client can move a sweep even from feet away from the reader. Further, in addition to being cost effective due to it's easy and less massive production cost, RFID alleviates the scope of malpractices and errors. One more highlight note here is that this framework can additionally be reinforced by presenting parts like a finger impression scanner assuming the need emerges. Likewise the extent of this RFID based framework can be reached out to a scope of tasks like library the board, stopping security the executives, and so forth.

References

1. Weis, S.A.: RFID (radio frequency identification): principles and applications. System **2**(3), 1–23 (2007)
2. Chou, T.S., Liu, J.W.S.: Design and implementation of RFID-based object locator. In: RFID, pp. 86–93. IEEE (2007)
3. Bonde, M.J., Rane, K.P.: Radio frequency identification (RFID) misplaced objects. Int. J. Electron. Electr. Eng. **7**(7), 657–662 (2014)
4. Chawla, K., Robins, G.: An RFID-based object localisation framework. Int. J. Radio Freq. Identif. Technol. Appl. **3**(1–2), 2–30 (2011)
5. Vogt, H.: Efficient object identification with passive RFID tags. In: Proceedings of the Pervasive, pp. 98–113 (2002)
6. Symonds, J., Parry, D., Briggs, J.: An RFID-based system for assisted living: challenges and solutions. Stud. Health Technol. Inform. **4**, 127–138 (2007)
7. Juels, A.: RFID security and privacy: A research survey. IEEE J. Sel. Areas Commun. **24**(2), 381–394 (2006)
8. Zhao, Y., Liu, Y., Ni, L.M.: VIRE: active RFID-based localization usingvirtual reference elimination. In: Proceedings of ICPP, pp. 56–56. IEEE (2007)
9. Chen, L.C., Sheu, R.K., Lu, H.C., Lo, W.T., Chu, Y.P.: Object finding system based on RFID technology. In: Shen, H.T., Li, J., Li, M., Ni, J., Wang, W. (eds.) APWeb 2006. LNCS, vol. 3842, pp. 383–396. Springer, Heidelberg (2006). https://doi.org/10.1007/11610496_51
10. Symonds, J., et al. (eds.): Auto-Identification and Ubiquitous Computing Applications. IGI Global (2009). https://doi.org/10.4018/978-1-60566-298-5
11. Bonde, M., Rane, K.: Radio frequency identification (RFID) misplaced objects. Int. J. Electron. Electron. Eng. **7**, 657–662 (2014)
12. Ajay, J., Aman, A., Arpit, S., Poonam, J.: RFID based attendance system. Int. J. Mod. Trends Sci. Technol. **07**(01) 40–43 (2021)
13. Chou, T.S., Liu, J.W.S.: Design and implementation of RFID-based object locator. In: 2007 IEEE International Conference on RFID, pp. 86–93 (2007). https://doi.org/10.1109/RFID.2007.346154
14. Chou, T.S., Liu, J.W.S.: Design and implementation of RFID-based object locators. In: IEEE International Conference on RFID (2007). ieeexplore.ieee.org
15. Reshma, A., Rajathi, G.M.: Voice based navigation system for visually impaired people using RFID tag. In: Hemanth, J., Fernando, X., Lafata, P., Baig, Z. (eds.) ICICI 2018. LNDECT, vol. 26, pp. 1557–1564. Springer, Cham (2019). https://doi.org/10.1007/978-3-030-03146-6_182

Depth Estimation and Navigation Route Planning for Mobile Robots Based on Stereo Camera

Ajay Kumar Kushwaha[1](✉), Supriya M. Khatavkar[1], Dhanashri Milind Biradar[2], and Prashant A. Chougule[1]

[1] Bharati Vidyapeeth (Deemed to be University) College of Engineering, Pune, India
akkushwaha@bvucoep.edu.in

[2] Sharad Institute of Technology College of Engineering, Yadrav, Ichalkaranji, India

Abstract. The extraction of three dimensional (3D) information from digital pictures, such as those acquired by a CCD camera, is known as computer stereo vision. Examining the relative locations of things in the two panels allows 3D information to be derived by comparing information about a scene from two viewpoints. We employed two cameras in the suggested system to identify obstructions in front of the autonomous vehicle or robot using the disparity idea. The absolute difference between the two pictures is calculated and utilized to regulate the motion of the vehicle/robot. Edge pixels retrieved from two pictures are matched, and a dense disparity map is generated by filling in the gaps between two consecutive edge pixels. A system has been proposed that permits the identification of an ideal path in real time between a starting point and a desired position in a congested environment using stereo vision, followed by a path planning algorithm and navigation enabling easy vehicle/robot traversal.

Keywords: Depth estimation · Disparity map · Stereo vision · Autonomous vehicles · Robot · Block matching · Sum of absolute difference (SAD)

1 Introduction

Recently, technology is steadily moving toward automation with little human intervention and reaching ease of operation in a wide range of disciplines. Since autonomous robots can automate a variety of labor-intensive occupations in the industrial environment and increase productivity, the number of robots used in manufacturing has rapidly increased, and this trend is predicted to continue in the future. Autonomous vehicles and robots may also do activities without the need for human intervention by combining hardware and the right algorithm to achieve the needed functionality in a range of contexts, which has led to their widespread use in several industries.

For navigation, a variety of sensors are used, such as ultrasonic sensors, laser scanners, Lidars, and others. This study looked at how to use stereo vision as the main sensor to create a consistent global map for an unstructured interior environment in real time,

N. Gupta et al. (Eds.): IC4S 2022, LNICST 472, pp. 180–191, 2023.
https://doi.org/10.1007/978-3-031-28975-0_15

which will aid in a robot's autonomous movement. Due to its numerous significances in several applications, such as Industrial Automation, ADAS, and robots, visual perception of an unknown environment has become essential [1]. Localization of the robot or vehicle, as well as sufficient map and path planning, are essential for autonomous traversal, with path planning being critical in implementing an ideal accident-free path forward. Precedent to these activities, we performed object recognition and range estimation for every item via the vision configuration.

Any sort of uncertainty perceived in a world model should be dealt with by the path planning method, which should also be capable of reducing the impact of objects on the robot when it is traversing a small space. These frequently use different map representations as their foundation, including metric maps, topological maps, hybrid maps, and others [2]. For these applications, several object recognition and classification algorithms built on stereo matching approaches [3], classic image processing techniques as well as deep learning-based methods [4] have been developed. We have used computer vision for this.

Stereoscopy is a technology for capturing and displaying 3D pictures by merging two or more photographs taken from slightly different perspectives, it may create the illusion of depth. There are two ways to capture stereoscopic images: with specialist two-lens stereo cameras or with systems that combine two single-lens cameras. The distance between the camera and the object of interest in the picture can be calculated using stereoscopic images. One of the most important qualities of any autonomous ground vehicle is its ability to discriminate between obstacles and how reliable and complete it is the perception of the environment.

After identifying an item and measuring distance from it, a system of stereo vision provides a pair of stereo images that may be used to measure distance. Avoidance performed by any of the controlling devices upon receiving the detection decision from the stereo system [5–8]. To grasp the different techniques published in the literature in recent years, a taxonomy of stereo matching methods is devised.

2 Methodology

Figure 1 shows the block diagram of the proposed system for depth estimation and navigation path planning using stereo vision, which is briefly described in this section.

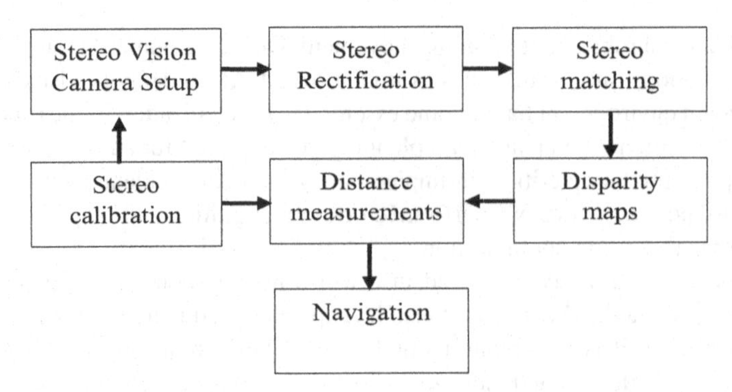

Fig. 1. Proposed diagram for depth estimates and navigation path planning

2.1 Stereo Calibration

When working with stereoscopy, it is critical to calibrate the cameras and get the necessary intrinsic and extrinsic characteristics. This is an example of a pinhole camera. A scene view is created by employing a perspective transformation to project 3D points into the picture plane.

$$s\begin{bmatrix} u \\ v \\ 1 \end{bmatrix} = \begin{bmatrix} f_x & 0 & c_x \\ 0 & f_y & c_y \\ 0 & 0 & 1 \end{bmatrix} \begin{bmatrix} r_{11} & r_{12} & r_{13} & t_1 \\ r_{21} & r_{22} & r_{23} & t_2 \\ r_{31} & r_{32} & r_{33} & t_3 \end{bmatrix} \begin{bmatrix} X \\ Y \\ Z \\ 1 \end{bmatrix} \tag{1}$$

When we calibrate the camera, we get certain precise numbers that we may use to estimate distances in length units rather than pixels. The purpose of calibration is to determine the intrinsic and extrinsic characteristics of the camera. A chessboard pattern is a common method for calibrating a camera. The chessboard patterns as shown in Fig. 2. The chessboard design is easy to calibrate since it is flat, so there are no depth problems, and it is easier to extract corner points because they are well defined. The

Fig. 2. Chess board patterns

corners are all on the same line. To get a better calibration, a variety of positions are employed [9–11].

Stereo calibration is identical to single camera calibration; however, it includes more stages and provides all intrinsic and extrinsic characteristics. We must define two image vectors and locate the chessboard in each photo.

2.2 Stereo Rectification

For rectification of frames received from stereo cameras, intrinsic and extrinsic properties discovered during calibration are employed. Due to the imaging principle of the camera and the construction of the device, the left and right pictures acquired by the binocular vision system normally contain some image distortion. As a result, the identical pixel in the left and right photos may not be on the same pole line. This will make subsequent stereo matching more difficult, resulting in increased time consumption and mismatch. To increase the accuracy of stereo matching, the picture must be corrected to obtain precise co-plane and line alignment of the left and right images in theory. Calibration and rectification of stereo images as shown in Fig. 3.

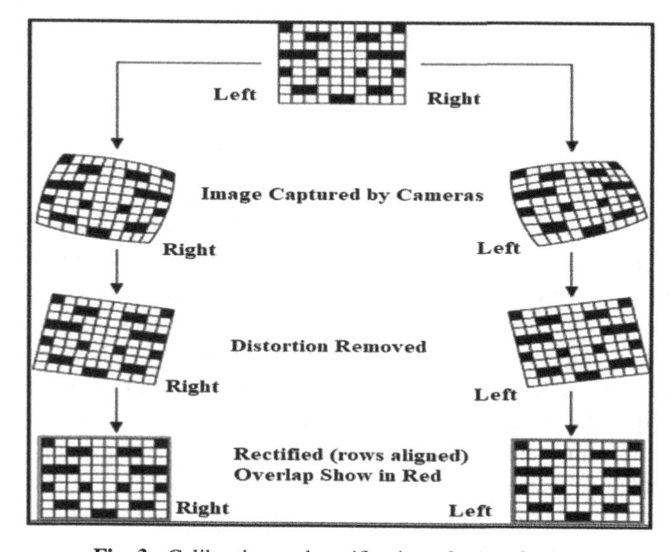

Fig. 3. Calibration and rectification of stereo images

2.3 Stereo Synchronization

Many stereo synchronization algorithms have been presented during the last two decades. All approaches are classified as sparse stereo or dense stereo matching. All approaches are classified as direct matching, custom-built filters, or network learning models. Until date, the majority prominent categorization approach has been global and local. One of the most significant responsibilities of a machine vision system is calculating the distance between distinct points or primitives in a scenario and the cameras location. The most common method for obtaining depth information from intensity images is to use two synchronized camera signals, from which the disparity maps, also known as depth images, are created by matching the two stereo images point by point [3–9, 11, 12].

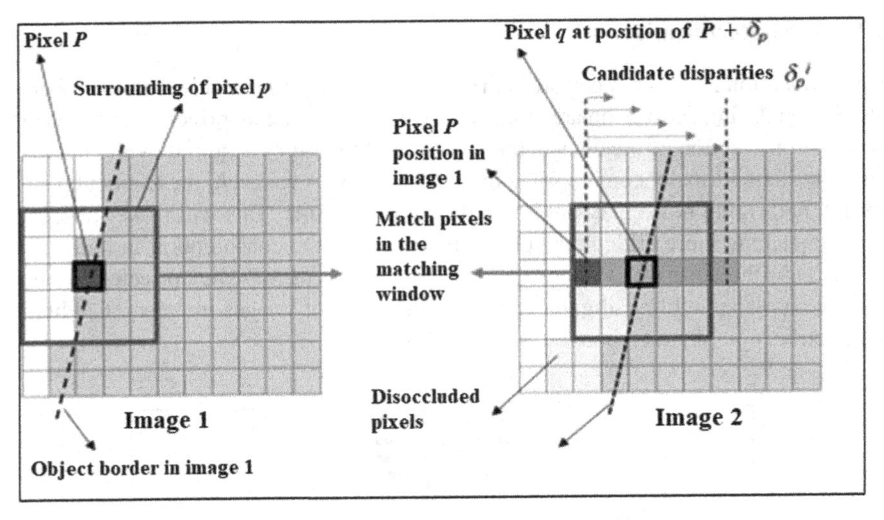

Fig. 4. Matching windows in stereo matching

To calculate the pixel p disparity, local stereo matching compares a pixel's surroundings in the left picture to significantly translated regions in the right image. The processing of each individual pixel, which ignores the context of the complete image, results in noisy disparity images. Areas of non-textured images that are impacted by input noise of any kind are especially susceptible to this (e.g., light gleaming, slightly varying colours over adjoining camera views, etc.). In Fig. 4, the surrounding areas of pixels p and q in the left and right images are compared using local stereo matching algorithms, with q translated across a candidate disparity p in comparison to p. N = 256 and 65,536 potential disparities are assessed for each pixel p in 8-bit and 16-bit depth maps, respectively, and the candidate disparity with the lowest matching cost is given to pixel p. [8]. On the other hand, accurate, local stereo matching algorithms pay close attention to the matching window form, lining up the edges with object borders.

2.4 Disparity

Disparity is the measured parallel shift in pixel-coordinates between the positions of a particular object in a pair of stereo pictures. By identifying the object in both the left and right photos, disparity is calculated. $d = xl - xr$, where xl and xr stand for the item's parallel locations in the left and right images, respectively, and d represents the disparity. Items close to the camera will be in a different position than those farther away. The relationship between disparity and depth allows us to calculate the actual separation between two objects. Figure 5 shows two images taken using a stereo camera setup along with the distance between the objects. Each image has a different object in two dimensions. The distance between the cameras and the object can be measured by calculating the disparity. Distance and disparity are inversely connected.

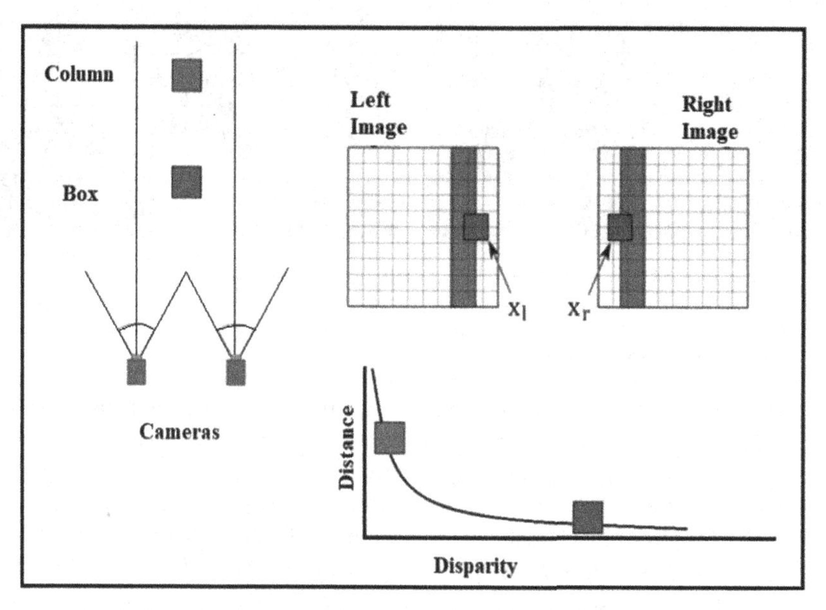

Fig. 5. The pair of photographs taken and the disparity between the objects

2.5 Disparity Map

A disparity map is a 2D matrix in which each quantity represents the pixel disparity value by a single value. By visualising each number as a coloured pixel, the disparity map can be displayed as a grayscale image. High disparity levels produce brighter pixels, whilst low disparity values produce darker pixels, as shown in the image. The disparity map is shown as a grayscale image in Fig. 6. Dark features are farther away from the cameras than bright features. Features that are too far away from the camera to be associated with are represented by the black patches.

Fig. 6. Disparity map for stereo pairs

2.6 Stereo Correspondence

The correspondence between two corrected stereo pictures is referred to as stereo correspondence. A disparity map is produced by computing the difference between the features of the left and right images. Stereo Block Matching to compute stereo correspondence, Python and OpenCV create a quick and efficient stereo block-matching technique. The semi-global block matching-method is the function that was implemented. The photos are first gone through excluding process to improve quality and make it easier to locate characteristics. In order to identify characteristics that complement those of the left and right corrected images, the approach iterates between the images using a SAD window (sum of absolute differences). The algorithm looks through the equivalent rows in the right image to find a match for each trait found in the left picture. The algorithms use the SAD window to match blocks of pixels rather than individual pixels. By doing this, each batch of photos will process more quickly, which is beneficial for real-time applications. Post filtering is used to prevent bad similarity matches from occurring. The class Stereo SGBM in OpenCV offers this capability. The disparity function in Python uses the same method. Consequently, a disparity map is produced.

2.7 Total Absolute Differences

In digital image processing, the sum of absolute differences (SAD) is a metric for picture block similarity. It is calculated as the sum of the differences between each pixel in the native block and its corresponding pixel in the contrast block. The block matching motion estimation subsystem uses the SAD method continually. The macro block uses the SAD technique to calculate the definite differences between the picture's (Template image) and its matching pixels' (Search image) pixels, and these differences are then averaged to produce the similarity block. The only arithmetic operations used in the procedure are addition and shifting. The SAD approach is particularly the quickest and is extensively used in block motion evaluation and object discovery due to its simplicity. It performs independent computation checks on each pixel in the block, making the implementation process simpler and more parallel. This study presents the 4 × 4, 8 × 8 SAD technique for video compression motion estimation. The architecture can conduct full locomotion searches on essential manifold of 4 × 4 and 8 × 8 block dimension. Figures 7 and 8 depict the diagrams of blocks of 8 × 8 SAD and Ladder of SAD respectively.

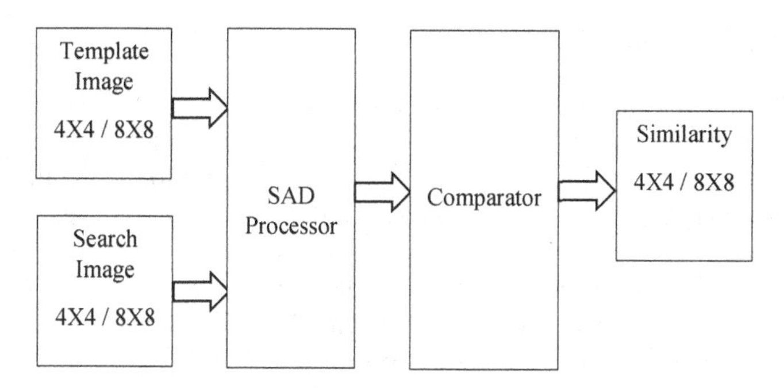

Fig. 7. Diagram of 8 × 8 SAD block

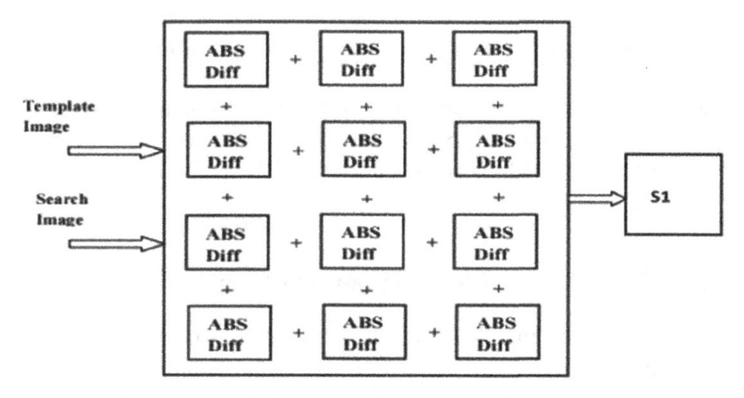

Fig. 8. Ladder of SAD block

Representative stages elaborated in completely parallel SAD architecture are:

- Carry out complete difference of all the pixels (of a block of video).
- Perform addition of all the complete dissimilarity.
- Pick block with smallest contrast value.

The sum of absolute differences (SAD) approach is an easy way to determine how similar template image T and sub-images in source picture S are to one another. The absolute difference between each pixel in T and its corresponding pixel in the sub-images being compared in S is determined. These distinctions are added together to form a basic similarity measure. Assume a 2-D m * n template, $T(x, y)$ is to be matched inside a $S(x, y)$ source image of size $p \times q$, where $(p > m \; and \; q > n)(x, y)$" [7]. The

SAD distance is determined for each pixel position (x, y) in the picture as follows:

$$SAD(x, y) = \sum_{k=0}^{(m-1)} \sum_{l=0}^{(n-1)} |S(x + k, y + l) - T(k, l)| \tag{2}$$

2.8 Distance Evolution

Range evaluation via stereovision is commonly utilized since, for a given environment, two separate perspectives are produced, which aids in determining depth values. Due to the disparity values obtained being inversely proportional to the depth of a certain scene, disparity maps are utilized to estimate the depth of a given scene. The bounding boxes produced by object detection were employed to estimate range for the detected items, with the centroid of the bounding box designating the location of the object in the environment at that precise moment [4]. The following are the distance estimate equations that were utilized based on this information:

$$\text{D} = \frac{\text{baseline} \times \text{focal length}}{\text{disparity value}} \tag{3}$$

$$D = 562.44 \times d^3 + 1426.83 \times d^2 + 1300.22 \times d - 494.35 \tag{4}$$

Here, d constitute of disparity values acquired from SAD method. This equation was created using ground truth data of noticed object detection interval and disparity values, where 'd' represents observed disparity values. These two equations were obtained to calculate the range of objects that were identified.

2.9 Physical Mapping and Navigation Route Planning

Building a map and improving it at orderly time spans that are synced with the robot's movement is a critical task for navigation. Mapping an unfamiliar area aid in determining the position of the robot in relation to its surroundings. This approach aids in the estimation of landmarks, which aids in the route planning process of the robot. The suggested technology performs mapping using vision, with items classified as barriers serving as markers for the robot. "To map the environment in our scenario, we use occupancy grid maps, which are made up of discretized cells that each reflect the occupancy of a certain obstacle. These assist in locating nearby open spaces that can be used for safe movement or to accomplish a particular objective. Each grid cell is given an integer value that describes its state, much like an occupancy grid map (empty or occupied). Obstacle-occupied cells are given a high integer value, i.e., 1, whereas empty cells are given a value of 0" [4]. A typical illustration of an occupancy grid map is shown in Table 1.

Following the completion of the techniques outlined in the preceding segments, produced data is received and provide to route planning methods, which generate control points that the robot must act in accordance with in order to get to the target location. These waypoints were then utilized to provide actuation orders to the robot, allowing it to move securely.

Table 1. Occupancy grid map

1	1	1	0	0	0	0	0
1	1	1	0	0	0	0	0
1	1	1	0	0	0	0	0
0	0	0	0	0	0	0	0
0	0	0	0	0	0	0	0
0	0	0	0	0	0	0	0
1	0	0	1	1	1	1	1
1	0	0	0	1	1	1	1
1	0	0	0	1	1	1	1
1	0	0	0	1	1	1	1

3 Result and Discussion

It is depended on the methods mentioned in preceding segments, a network map was created to indicate the habitation of every network by the entity and was supplied to the planner to generate the course that the robot travelled. The range estimates for each identified barrier were calculated using Eqs. (2) and (3), after which some observations were taken, revealing that the suggested interval-disparity mapping equation, Eq. 3, proved to be more precise than Eq. 2.

Figure 9 depicts the stereovision system's output GUI. A initial camera image is shown in GUI Image 1. An image from the second camera is displayed in Image 2. The

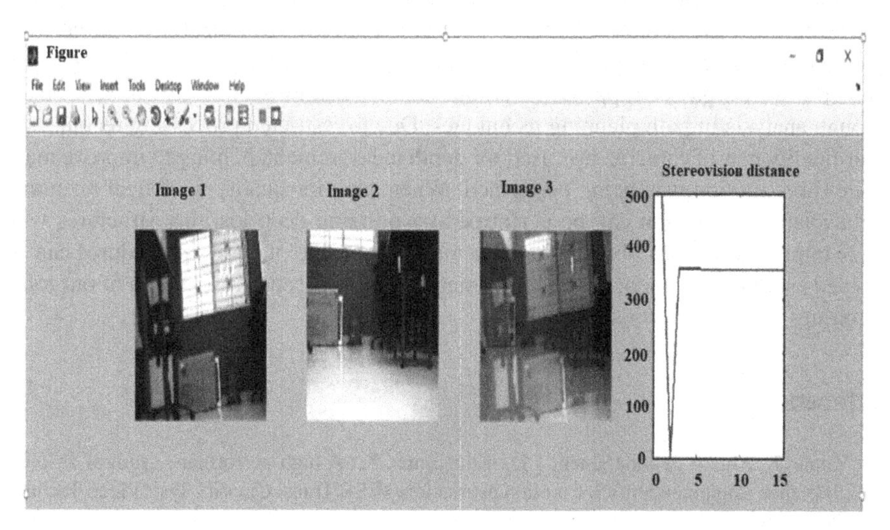

Fig. 9. Output GUI of stereovision system

composite image of images 1 and 2 is seen in image 3. The distance travelled by the object is displayed on a stereovision distance graph.

Table 2. Comparison for range estimation with equations

Ground truth (cm)	Equation 2 (cm)	Equation 3 (cm)
31	41	31
61	56	64
91	61	93
121	64	124
151	79	148
181	89	180
211	121	208
241	127	241
271	146	274

Table 2 mention comparison for range estimation with equations. The suggested method, in conjunction with the distance-disparity mapping equation, proved to be efficient in presented scenario.

4 Conclusion

The procedures mentioned here for autonomous route planning of a robot/vehicle in an interior territory are conducted in real-time using the provided object perception structure and the suggested distance estimate approach. The performance of the suggested method as well-organized approach for route planning in limited contexts was determined through analysis of path planning techniques. Due to restricted processing recourses, a high-fidelity pair of cameras was used for depth measurement, helping to improve maps while robot motion was being performed. When superior quality analytical resources are available, estimation can be performed by utilizing deep learning structures with maps improved using probabilistic route-maps, and route planning procedures can be evolved using more systematic planning methods for vigorous presentation of our robot in settings with severe constraints.

References

1. Vanne, J., Aho, E., Hamalainen, T.D., Kuusilinna, K.: A high-performance sum of absolute difference implementation for motion estimation. IEEE Trans. Circuits Syst. Video Technol. **16**, 876–883 (2006)
2. Salzman, O., Halperin, D.: Asymptotically near-optimal RRT for fast, high quality motion planning. IEEE Trans. Rob. **32**, 473–483 (2016)

3. Bansal, V., Balasubramanian, K., Natarajan, P.: Obstacle avoidance using stereo vision and depth maps for visual aid devices. SN Appl. Sci. **2**(6), 1–17 (2020). https://doi.org/10.1007/s42452-020-2815-z
4. Phan, R., Androutsos, D.: Robust semi-automatic depth map generation in unconstrained images and video sequences for 2D to stereoscopic 3D conversion. IEEE Trans. Multimed. **16**, 122–136 (2014)
5. Ren, S., He, K., Girshick, R., Sun, J.: Faster R-CNN: towards real-time object detection with region proposal networks. IEEE Trans. Pattern Anal. Mach. Intell. **39**, 1137–1149 (2017)
6. Chen, L., Fan, L., Xie, G., Huang, K., Nüchter, A.: Moving-object detection from consecutive stereo pairs using slanted plane smoothing. IEEE Trans. Intell. Transp. Syst. **18**, 3093–3102 (2017)
7. Alsaade, A.: Fast and accurate template matching algorithm based on image pyramid and sum of absolute difference similarity measure. Res. J. Inf. Technol. **4**, 204–211 (2012)
8. Stankiewicz, O., Lafruit, G., Domański, M.: Multiview video: acquisition, processing, compression, and virtual view rendering. In: Academic Press Library in Signal Processing, vol. 6, pp. 3–74 (2018)
9. Murray, D., Little, J.J.: Using real-time stereo vision for mobile robot navigation. Auton. Robot. **8**, 161–171 (2000). https://doi.org/10.1023/A:1008987612352
10. Yurtsever, E., Lambert, J., Carballo, A., Takeda, K.: A survey of autonomous driving: common practices and emerging technologies. IEEE Access **8**, 58443–58469 (2020). https://doi.org/10.1109/ACCESS.2020.2983149
11. Bányai, T., Cservenák, Á.: Logistics and mechatronics related research in mobile robot-based material handling. In: Jármai, K., Cservenák, Á. (eds.) VAE 2022, pp. 428–443. Springer, Cham (2023). https://doi.org/10.1007/978-3-031-15211-5_36
12. Kushwaha, A.K., Kumar, A.: Sinusoidal oscillator realization using band-pass filter. J. Inst. Eng. (India): Series B **100**, 499–508 (2019)

Water Level Forecasting in Reservoirs Using Time Series Analysis – Auto ARIMA Model

Avinash Reddy Kovvuri, Padma Jyothi Uppalapati(✉), Sridevi Bonthu, and Narasimha Rao Kandula

Vishnu Institute of Technology, Bhimavaram, A.P, India
20pa1a0585@vishnu.edu.in, padmajyothi64@gmail.com

Abstract. Forecasting the upcoming water level of a dam or reservoir is the goal of water level forecasting in reservoirs. In order to predict the water level of the dam or reservoir for the subsequent consecutive time interval, this paper proposes a method based on the ARIMA (Auto Regressive Integrated Moving Averages) machine learning model, which fed on historical data of water levels with respect to consecutive time intervals. Additionally, the anticipated output, whether it be in TMC or MFTC units, is depending on the data that is given. The model's performance is further examined in the study using certain machine learning metrics.

Keywords: Water Level Forecasting · ARIMA · Time Series Analysis · Auto ARIMA

1 Introduction

Water Level Forecasting in Reservoirs is to forecast the future outflow of city or country reservoirs. The purpose of predictions or forecasts is to make ourselves ready to meet the future needs and to make the ruling bodies aware of the future trends of any given commodity. This solution or analysis is used as a basis for any government bodies to make any substitutions to the insufficient levels of water for the people. Now a days the techniques of machine learning are making a huge impact in society or the market by getting an analysis on the future needs. And the machine learning model used in this is ARIMA which is a Time Series Analysis model based on the single variable data which varies with respect to the time [1].

The Word Time Series Analysis means by analysing a sequence of data collected over an interval of time [2]. ARIMA or Auto Regressive Integrated Moving averages which is a combination of AR model (Auto Regressive) and MA (Moving Averages) model with respect to differencing (Stationarizing data) [3].

The incident laid foundation for the idea - Forecasting water levels was due to the vast growth of the capital city of Tamil Nadu state in India named Chennai

N. Gupta et al. (Eds.): IC4S 2022, LNICST 472, pp. 192–200, 2023.
https://doi.org/10.1007/978-3-031-28975-0_16

from 1893 to 2017, areas of the surrounding floodplain, along with its lakes and ponds had disappeared. This leads to the decrease of Chennai's water bodies from 12.6 km^2 to about 3.2. And, finally the water crisis in 2019 was declared as "Day Zero" by city officials on 19th June. The people along with the government face a huge problem with the sudden lack of water for the entire city. In order to solve the above, forecasting the water levels through Time Series Analysis Technique in machine learning will –

- The forecasted data used by the government to fulfil the needs of the people in Chennai city regarding water whether there is any chance of occurrence of water crises by comparing it with the population.
- With forecasted data, the officials can also evacuate the people of Chennai city.
- Prevention measures for any issues related to floods were taken into consideration.

2 Chennai City's Reservoirs Data

Actually the city Chennai has a source of 4 main reservoirs named Poondi, Cholavaram, Redhills, Chembarambakkam. The source for these 4 dams was rainfall water. These 4 dams together with addition of extra rainfall water adding up to another reservoir as a source for people of Chennai city. The water in the final reservoir plays a key role for the people of Chennai city in order to make decisions on the usage of water. According to the statistics on the day zero in chennai in 2019, Due to the drain of reservoirs named poondi from 3,231 to 22, cholavaram from 1,081 to 0 and redhills, chembarambakkam from 3300 and 3645 to 0, being india's fourth largest city, needs about 800 million litres of water daily but the public water board has been able to supply 525 million litres which impacts the people along with government officials of the state.

3 Time Series Analysis

Time Series Analysis can be applied on the time-series data which is a sequence of data noted or stored with respect to specific intervals of time in chronological order [4]. This level of forecasting has a huge impact on economic, commercial and other financial aspects of business and life. This type of analysis plays a major role in situations of natural calamities.

Features of Time-series are:

- Trend
- Seasonality
- Cyclicality
- White noise

It's mandatory that the data must be stationary if we are using time-series forecasting. Stationarity in the data indicated that the distribution of the data doesn't change with the time. We explain it with the statistical measure such as trend, variance, autocorrelation to remain constant. In general, stationary data is said to have the following three properties:

- Trend = zero
- Variance = constant
- Autocorrelation = constant

The trend may be upward, downward or constant. For a stationary time-series the trend must be constant. In Simple words, Variance is of average distance of the specified data with respect to certain time interval from the zero line in the graph if we draw the varying time quantity with respect to time.

The Variance must be constant for a time-series data to be Stationary. Autocorrelation is nothing but how each value or observation in time-series data relates to its neighbours and it must be constant for a data to be stationary.

If the data is non-stationary and if fed to the model, the final results may be highly inaccurate. The conversion of non-stationary to stationary by using techniques of Differencing or log of the series. Most commonly used method is differencing.

4 ARIMA

ARIMA or Auto Regressive Integrated Moving Averages is a statistical analysis model there on top of all other time-series models which feed on time series data to either get clear insight about the data or forecast future values. It is a combination of AR and MA with respect to differencing (Stationarizing data) [5].

4.1 Auto Regression (AR)

In simple words, it is a model which regresses on its own lagged, or previous values by changing variable [6]. The Eq. 1 shows the equation represents the AR model. Epsilon-t is white noise or term that represents shock-term at that particular time. White noise is a series of measurements in which each value is uncorrelated with previous values.

$$y_t = c + \phi_1 y_{t-1} + \phi_2 y_{t-2} + ... + \phi_p y_{t-p} + \epsilon_t \tag{1}$$

4.2 Integrated (I)

It refers to some techniques where the actual data was differenced to convert it to stationary and finally new data is replaced with old non-stationary one.

4.3 Moving Averages (MA)

It represents an equation which regresses values of time series against previous shock values of the same time series as of Eq. 2.

$$y_t = c + \epsilon_t + \theta_1 \epsilon_{t-1} + \theta_2 \epsilon_{t-2} + ... + \theta_p \epsilon_{t-p} \tag{2}$$

The Eqs. – 1, 2 represents the equation which shows the combination of rela tion of specified data to its previous with respect to time with an order of q. The three components involved in ARIMA which are of integers whose functionality defined as:

- p: Integer value, which decides the number of values to be regressed is used in the AR model.
- d: Integer value, which is the number of times the differencing tech-nique applied on data in order to convert to stationary.
- q: Integer value, which decides the number of shock terms with respect to time, is used in the MA model.

$$y_t = \mu + \sum_{i=1}^{p} a_i y_{t-i} + \sum_{i=1}^{q} b_i \epsilon_{t-i} + \epsilon_t \tag{3}$$

$$\epsilon_t = \sqrt{\sigma_t Z_t}, \sigma^2 = w + \sum_{i=1}^{p} \alpha_i \epsilon_{t^2-i} + \sum_{i=1}^{q} \beta_i \sigma_{t^2-i} \tag{4}$$

The Eq.-3, 4 shows the equation of the ARMA model which is a time-series model, regressed on previous values and previous shock terms]cite7. After forecasting through the ARIMA model, the values are stationary but we need non-stationary (based on input). For that we actually convert by using the values of differencing and the technique used in it.

For **differencing**, we use cumulative sum. The count of differencing applied on the data to achieve stationarity before it is fed to the model will be the number of times the cumulative sum must be applied in order to achieve non-stationarity in the result.

For the **log method**, we use an exponential function. The number of times the log method is applied on the data to achieve stationarity before it is fed into the model will be the number of times the exponential function will be used to achieve nonstationarity in the result.

It's important to note that when we process the data, we tend to remove the seasonality (patterns repeated at regular intervals) in the data. Even after we remove the seasonality in the data, but still the data holds the seasonal properties, then such data can't be fed to the ARIMA model, such data can be satisfied using the SARIMA or the Seasonal ARIMA model.

5 Data Preparation

After collecting the data with respect to time i.e., the final reservoir which is a combi-nation of all the 4 dams in Chennai city. The data is stored in colaboratory notebook of Google in order to execute. The data is dated from 2014 and to the year 2019 and the sample of top 5 observations. First step needed to check before going to feed data to the model is stationarity. Can check stationarity through visually and various tests. Through Fig. 1 and the concept of Stationarity in ARIMA in above, the data is non - stationary. And a test named AD- Fuller used to verify the data used is stationary or non-stationary.

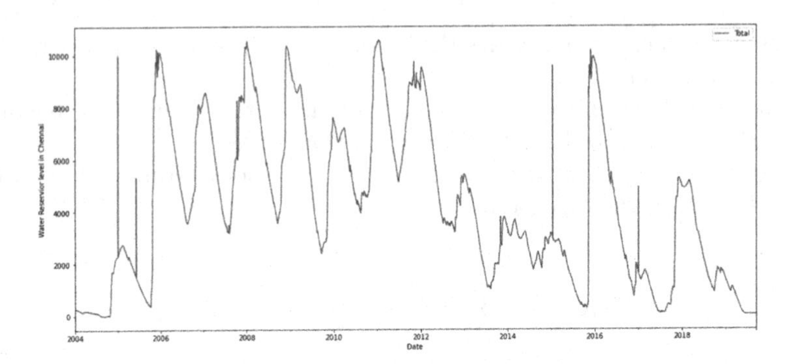

Fig. 1. Plotted Data

5.1 Ad-Fuller Test

Augmented dicky Fuller is one among the tests that are used to verify that the data used in the work is stationary or not. And is the most common test used in the processes which results in some values. If the obtained value is less than threshold then we say that data is stationary. Ad-Fuller test is performed by importing from "statsmodels.tsa.stattools". After performing the AD-FULLER Test also the result same as above stated from data visualization i.e., data is non-stationary. In order to convert data to stationary the technique of differencing or log is applied on the data and continues to repeat the same until the AD-FULLER Test results that data is sta-tionary and counts the number of time applied as the value of the term "d". Figure 2 shows the plotting of the data after converting it into stationary through differencing. And the next step is to train the ARIMA model.

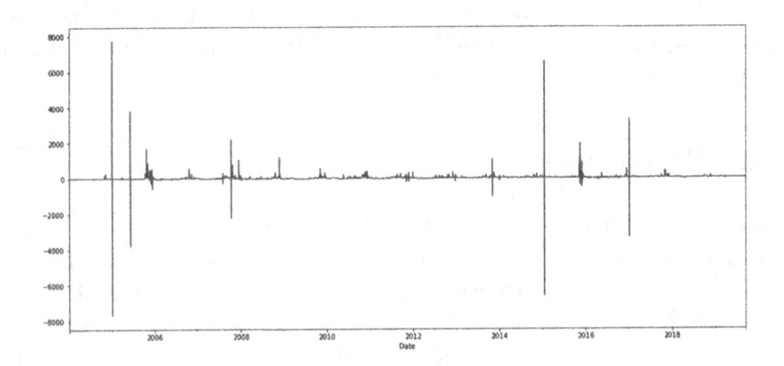

Fig. 2. Stationary Data

6 Training Arima Model

The data is ready to feed the model and the important thing is to find the parameters i.e., p, q and d. The d was also found while converting. Here the actual data i.e., nonstationary is fed along with the hyperparameters. The model itself will do the differencing through the parameter d. i.e., itself converting the data into stationary.

6.1 Experiments

The hyper-parameters were adjusted based on the result obtained through changing the values and the values which give the best scores of AIC and BIC (Minimum) are the best parameters [9]. The identification of p and q terms was done using Partial Autocorrelation and Autocorrelation in Fig. 3. After fixing the terms p, q, d the data is fed to the ARIMA Model along with the hyperparameters. And some consecutive data is left over to test the model. Finally the model is ready to predict the future values of the water levels of the Chennai city's reservoir.

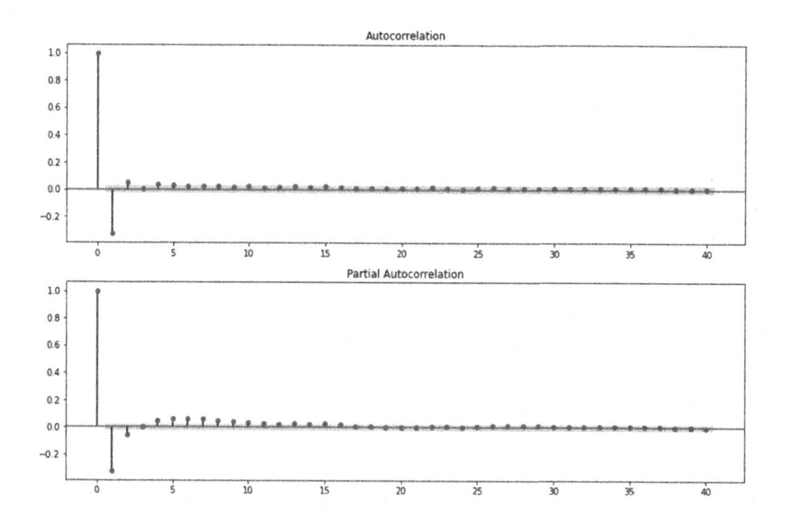

Fig. 3. Autocorrelation and Partial Autocorrelation

7 Auto ARIMA

For a while, we have been going through the process of manually fitting different models and deciding which one is best. So, we are going to automate the process. Usually, in the basic ARIMA model, we need to provide the p, d and q values which are essential. We use statistical techniques as above mentioned in order to generate these values by performing the difference to eliminate the non-stationary and plotting ACF and PACF graphs. In **Auto ARIMA**, the model itself will generate the optimal p, d and q values which would be suitable for the data set to provide better forecasting [10].

We automated the process of finding p, d and q values by using the Auto ARIMA model and data is fed to the model in order to predict future consecutive values [11].

8 Evaluation

Finally, we need an evaluation metric in order to know the performance of the model. So, we take 20 or 30% of the whole data as test data and compare it with the predicted data from our model in order to get a score of the model. Here the end consecutive data is separated from the training data and is used to test the model.

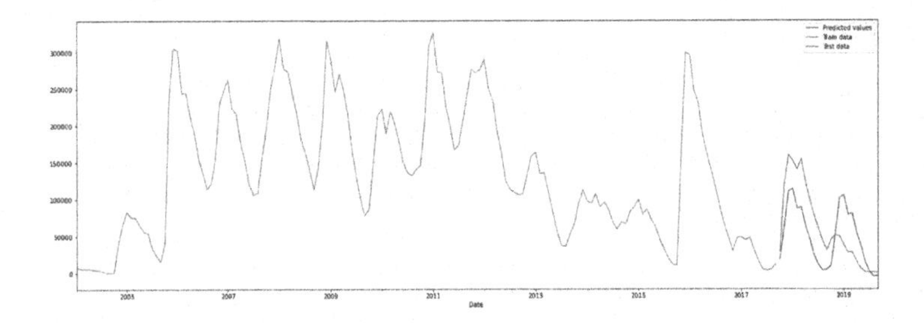

Fig. 4. Plot of Test, Predicted Data and Train Data

In Fig. 4 represents the graphical representation of the data available to us along with the differentiate between train, test data and data predicted by the model. The orange, blue and green colored lines in Fig. 4 indicates the train, predicted and test data.

The Fig. 5 indicates or plotted in order to get clear insight among predicted and actual data. The orange and blue lines in Fig. 5 indicates actual and predicted values. Figure 6 is plotted to showcase the fluctuations of current available and future pre-dicted data by plotting. The orange and blue line in the Fig. 6 indicates predicted fu-ture and total data. The evaluation metric used for the model is r2 score [12] and the obtained r2 score for the above model is 0.3284, which is greater than zero and is nearer to it. So, our model fits well and is able to predict the future data based on past outcomes.

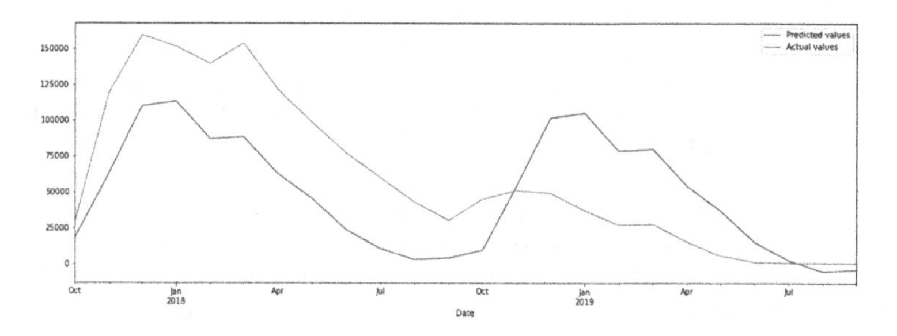

Fig. 5. Plot of Predicted and Actual Data

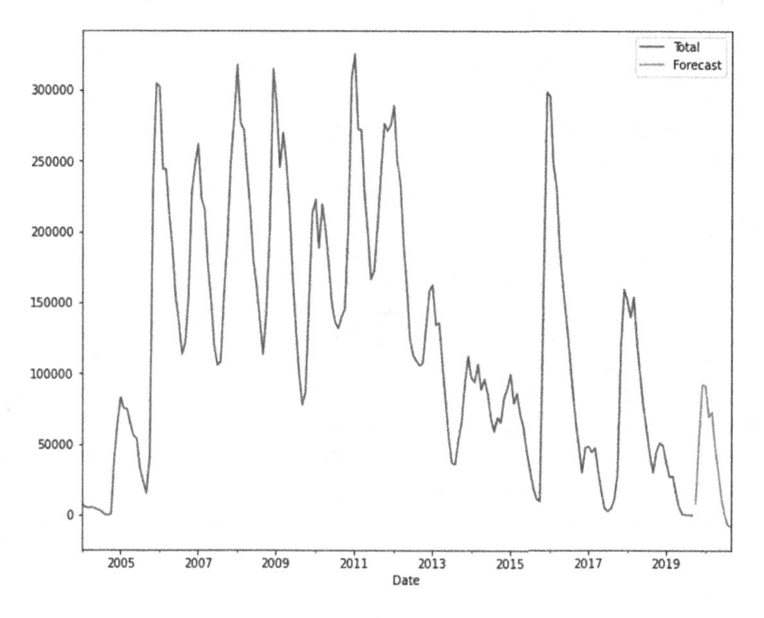

Fig. 6. Plot of Total and Future Data

9 Conclusion and Future Work

ARIMA Model is on top in time-series analysis in order to forecast upcoming data and we automated it with help of Auto ARIMA in order to increase the accuracy of the model. The proposed model is based on Time-series analysis and achieved nota-ble accuracy and good performance. As time-series analysis requires a high amount of data even for predicting for a shorter period of time. So, in order to make the model robust in terms of performance is to train the model on newly updated data at consequent time intervals. Finally the Auto ARIMA model was incorporated to know the future values of availability of water for Chennai capital city of Tamil Nadu in India.

References

1. Nguyen, X.H.: Combining statistical machine learning models with ARIMA for water level forecasting: the case of the Red river. Adv. Water Res. **142**, 103656 (2020)
2. Titolo, A.: Use of time-series NDWI to monitor emerging archaeological sites: case studies from Iraqi artificial reservoirs. Remote Sens. **13**(4), 786 (2021)
3. Wang, J., et al.: Reliable model of reservoir water quality prediction based on improved ARIMA method. Environ. Eng. Sci. **36**(9), 1041–1048 (2019)
4. Arvor, D., et al.: Monitoring thirty years of small water reservoirs proliferation in the southern Brazilian Amazon with Landsat time series. ISPRS J. Photogram. Remote Sens. **145**, 225–237 (2018)
5. Skariah, M., Suriyakala, C.D.: Forecasting reservoir inflow combining exponential smoothing, ARIMA, and LSTM models. Arab. J. Geosci. **15**(14), 1–11 (2022)
6. Huang, L., et al.: Evolutionary optimization assisted delayed deep cycle reservoir modeling method with its application to ship heave motion prediction. ISA Trans. **126**, 638–648 (2022)
7. Üneş, F., et al.: Estimating dam reservoir level fluctuations using datadriven techniques (2019)
8. Paparoditis, E., Politis, D.N.: The asymptotic size and power of the augmented Dickey-Fuller test for a unit root. Econ. Rev. **37**(9), 955–973 (2018)
9. Bai, Z., Choi, K.P., Fujikoshi, Y.: Consistency of AIC and BIC in estimating the number of significant components in high-dimensional principal component analysis. Ann. Stat. **46**(3), 1050–1076 (2018)
10. Yan, B., et al.: Flood risk analysis of reservoirs based on full-series ARIMA model under climate change. J. Hydrol., 127979 (2022)
11. Tegegne, G., Kim, Y.-O.: Representing inflow uncertainty for the development of monthly reservoir operations using genetic algorithms. J. Hydrol. **586**, 124876 (2020)
12. Ali, A., Bello, A.M., Raymond, J.: Machine learning algorithms for predicting reservoir porosity using stratigraphic dependent parameters. Glob. J. Comput. Sci. Technol. (2022)

Water Quality Monitoring Using Remote Control Boat

A. Prabhakara Rao[1](✉), K. V. S. H. Gayatri Sarman[1], Gopi Veera Pavan Kumar[1], and Sree Deekshitha Yerra[2]

[1] Department of ECE, Vishnu Institute of Technology, Bhimavaram, A.P, India
prabhakararao.a@vishnu.edu.in

[2] Department of CSE, Vishnu Institute of Technology, Bhimavaram, A.P, India

Abstract. Health of human beings, animals, and plants depends a lot on the quality of drinking water. There are many sources of drinking water such as lakes, canals, reservoirs, etc. Manually monitoring the water quality of these water bodies requires lot of effort as operators need to get in a boat with all the sensors and manually check the entire water body. In this work, to ease the water quality monitoring of water bodies, a remote control (RC) boat was used. This RC boat accompanied by different sensors was used to measure the PH level and turbidity level. When there is a deviation of water quality parameters from the standard values, the system will send an alert message to the remote user. Water quality was measured using two different sensors such as PH, temperature and turbidity sensors. PH sensor monitors the PH value of water and the turbidity sensor will find the suspended or unwanted particles in the water. The location from where the data collected from pH sensor, Turbidity sensor, and Temperature sensor was determined by using the GPS module. The measured values were displayed on the LCD. This data with respect to the location will be stored in the SD card and also the system will alert the user by sending a message with location, when the RC boat is about to sink and/or if the measured water quality is abnormal.

Keywords: Water quality monitoring · Remote control boat · PH sensor · Turbidity Sensor

1 Introduction

The health of animals and humans depends on the quality of their drinking water. There are several sources of drinking water such as rivers, reservoirs, lakes, etc. Water quality is extremely important to irrigation, fishing, and energy producing businesses. Water quality of these water bodies is important to be maintained at a specified level for water supply in urban houses and available water sources. However, it is not fundamentally secure for use in rural areas. In fact, despite the fact that it is the obligation of the government to ensure that its residents have access to clean water. Imperfectly the constant expansion of the population puts a burden on the infrastructure. The quality of drinking water plays a vital role in the lives of human beings and animals. Monitoring is defined

N. Gupta et al. (Eds.): IC4S 2022, LNICST 472, pp. 201–212, 2023.
https://doi.org/10.1007/978-3-031-28975-0_17

as the gathering of data at certain locations and intervals in order to produce information that may be utilized to guide present situations, establish trends, and so on. Sampling and laboratory procedures are two common approaches for determining water quality. These procedures, on the other hand, are inefficient and time-consuming, resulting in a delay in the detection of impurities and reactions to those pollutants in water. As a result, more efficient and effective water quality monitoring systems are needed. Water-dependent microbiological and physiochemical data can be used to do this. Turbidity, power of hydrogen pH, and temperature are some of the physiochemical properties of water [1, 2].

Instead of water sampling and laboratory testing, these characteristics are frequently assessed more economically and rapidly [3]. According to a study and research conducted by the United States Environmental Protection Agency [4], pollutants impact water parameters in precise ways that may be discovered, detected, and monitored using appropriate or specific water quality sensors. Commercially accessible probes and meters for water quality testing are available. These products analyze the parameters individually.

In this work, a low-cost, real-time, multi-sensor boat system that is specifically built for large-scale aquatic environments including rivers, reservoirs, and lakes is presented. PH, temperature, and turbidity sensors from WHO standards are used in the system. All sensor data is collected, evaluated, and communicated to the observer through a wireless communication system. The graphical user interface GUI approach is used to create and show these findings, complete with their readings and as well as their nominal ranges.

2 Methodology

In the proposed system, multiple components, modules like Arduino mega, GPS module, GSM modul, SD card module, battery, pH sensor, Temperature sensor, Turbidity sensor, water sensor, etc. were integrated to achieve the desired tasks. It is used to detect, monitor and keep track of records of the real instance parameters like pH, temperature and turbidity using the corresponding sensors like pH sensors, temperature sensors, and turbidity sensors to check the water quality [5]. A water sensor is used along with these sensors to detect if there is any water entering inside the remote control (RC) boat. The obtained data is processed in microcontroller and is sent to the user or costumer, as a message with real location and display on LCD display. Arduino mega 2560 is a development board based on the ATmega 2560 micro controller. It can be used and applied into work for lots of IOT projects and several other applications as this board gives a more view and add-ons for the projects that requires more GPIO pins and memory space. Arduino is the heart of our project. GPS Module contains tiny processors and antennas, which receives a certain location, along with timestamps, from a satellite through dedicated RF frequencies. Using this module, the RC boat can send real-time messages to the user. Global System for Mobile communication is shortly termed as GSM. Basically is designed and was used for wireless radiation monitoring through Short Messaging Services (SMS), mainly in transmitting the data as text SMS to a host server. Out of the box, the shield will work with the Arduino Uno. An SD card Module is a breakout board used for SD card processes such as reading and writing with a microcontroller, which allows communication with the memory card and write or read the information on them. The Arduino in our model can create a file in SD card to read, write and save data using the SD library. A battery is used in this device, to convert chemical energy into electrical energy. Here come the main components. PH sensor helps to measure the acidity or alkanity of the water with the value range between 0–14. pH stands for potential hydrogen. The basic principle of the pH sensor, or a pH meter, is to measure the concentration of hydrogen ions. To detect and measure the degree of hotness and coolness, and convert it into an electric signal, a temperature sensor is used. To measure the amount of light that is scattered by the suspended solids in water, turbidity sensors were used. With the use of LED light sources, turbidity sensors determine the level of particulate or suspended matter in water or other fluids. The vital component, water sensor is used to detect the presence of water, a leak, level, volume, etc. When Wi-Fi is enabled, the sensor can send out a notification to the homeowner through a Smartphone app. Besides these components, we used several other supporting components like LED, LCD (16*2), buzzer, jumper wires, motor driver modules (L298N), PCB (Printed Circuit Boards) and power supply circuits like resistor, voltage regulator, diodes, capacitors, analog joystick and etc.

In the proposed system, RC boat is accompanied by various types of sensors which are used to measure the pH level, turbidity level, temperature level. This will enable automatic monitoring of water quality in water bodies, more particularly in aquaculture [6].The RC boat is implemented [7, 8], integrating the GSM Module, GPS Module, Water sensor, LCD display, SD card Module, Temperature sensor, Turbidity sensor, pH sensor, LED and buzzer, to the Arduino Mega 2560. Block diagram of the proposed system is shown in Fig. 1.

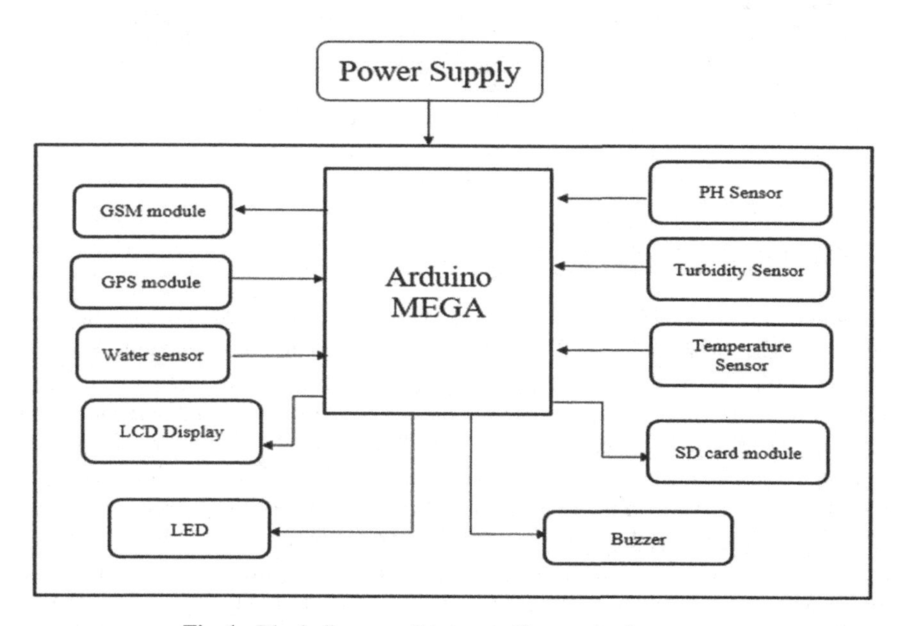

Fig. 1. Block diagram of water quality monitoring system

The power source is given to the device; the device comes to on condition. The commands are given through Arduino nano to the transmitter. The transmitter transmits the input signal to the receiver. The receiver, receives the input signals and decodes it into valid signals, and sends it to the Arduino nano, in which, several other components that are mentioned in above description. This process is a part and parcel of the device. Since, the operation of sensors and modules, is mainly is based on the input signals, which effects the final output or result.

Interfacing of Sensors, Modules (GSM, GPS, SD card), LED, LCD with Arduino MEGA by power supply in circuit diagram is as shown in Fig. 2.

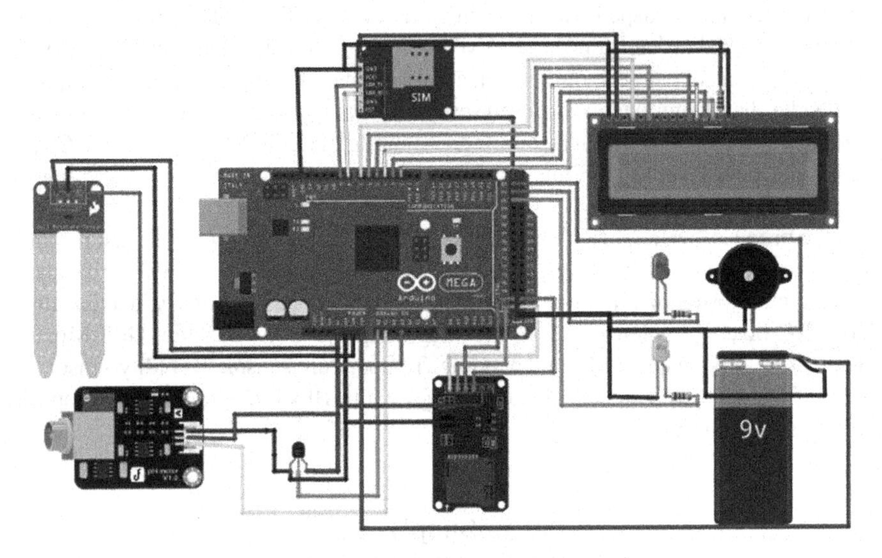

Fig. 2. Schematic of the proposed system

3 Working

The proposed project, GSM based Remote Control Boat (RC Boat) works as depicted in the flowchart shown in Fig. 3. There are various conditions and commands, operated in between the start and stop phase or on and off states or active and passive states. Let us discuss the operations performed in the active state, say when the device is on. Firstly, as the proposed project is remote control based, it obviously requires a remote to control the water quality analysis device. The movement of the boat, whether to continue using the boat, is based on the remote control commands. For the movement, if there is a command not to run the boat, then the process terminates. Else If the command is yes, the operations are executed step by step as follows. The microcontroller initializes the sensors and collects the detected, pH, temperature, turbidity values in real time from the sensors. These values were received by the microcontroller for further analysis and required action. If there is no deviation in the obtained data from the predefined values of pH, temperature and turbidity, corresponding to the regional water bodies, then the obtained data is stored in the SD card and the device displays those values on the display. If incase, there occurs a deviation in the obtained data, the device sends a real time message to the remote user with location. In both the cases the next proceeding step is continuing the data delivery and the process are terminated once after the user is noticed with the monitored status.

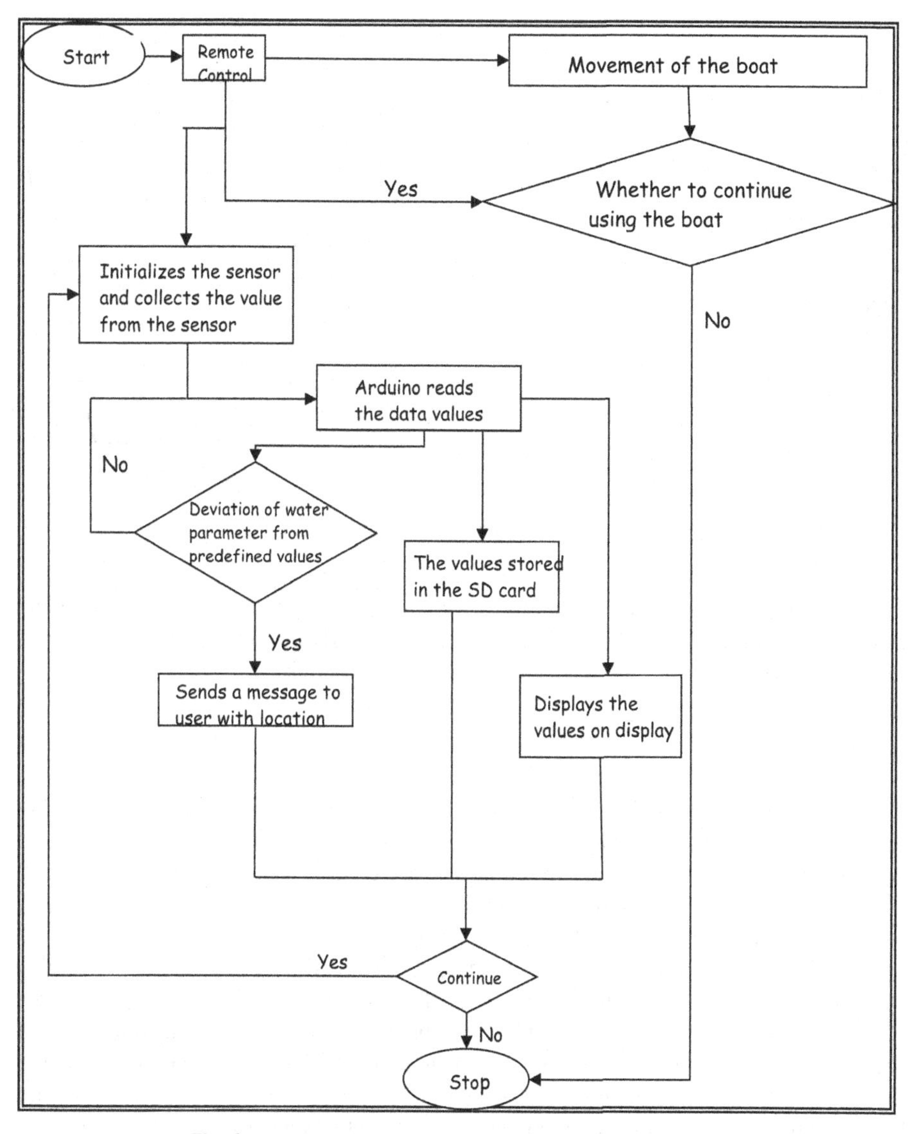

Fig. 3. Flow chart of the water quality monitoring system

The RC boat and remote controller block diagram is as shown in Fig. 4. In this system, Arduino Nano was used as a microcontroller for both remote controller and RC boat. Analog joystick and NRF24L01+ are connected to the Arduino nano in the remote controller. Power source for remote controller is given by using circuits and carbon zinc (HW) battery.

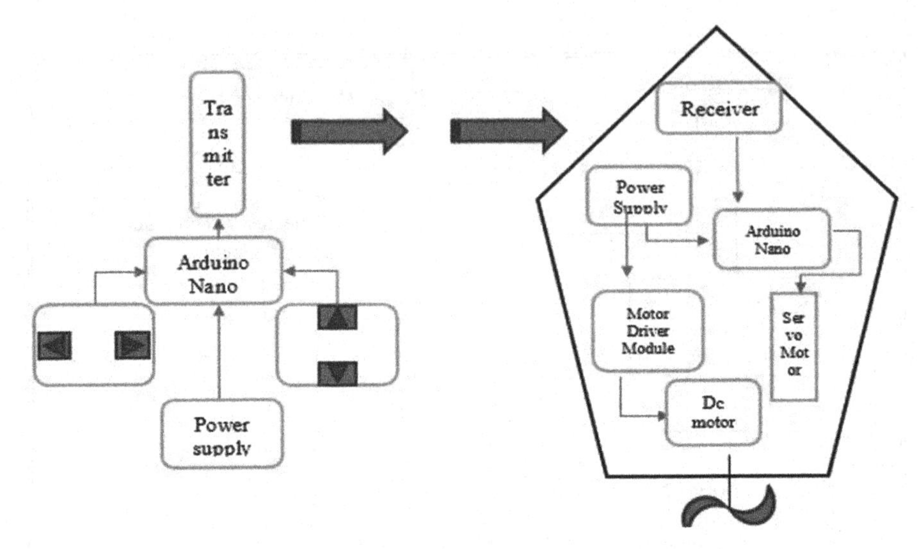

Fig. 4. Block diagram of remote controller and RC boat

NRF24L01, servo motor and motor driver module are connected to Arduino nano as input and output. DC motor is connected to motor driver module. Power source for remote controller is given by using circuits and Li-ion batteries.

4 Results

The water quality monitoring results were taken from different areas. The water quality parameters pH values, Turbidity values, temperature values and GPS coordinates [9] are collected by the Arduino and are stored in SD card module by using written command. It stores data for every 4 s. Further it will send message alerts through GSM and water, commands were displayed on LCD 16*02 display.

Water quality monitoring system was tested in some different scenarios. Also, this system can be used to monitor the water quality in remote areas [10]. The scenarios considered here are water in water tank of a building, lake in different timing. If pH values are in between 6.5 to 7.5 then water is neutral i.e., drinkable water, less than 6.5 then water having acidic nature i.e., unsafe to drink, greater than 7.5 then water having basic nature, i.e., likely to be contaminated with pollutants. If turbidity value is greater than 4 NTU then water is clean, in between 4 to 3.5 turbidity is present i.e., water is slightly clean, less than 3.5 NTU then turbidity is present i.e., water is not clean and it is impure.

The measurements depicted in Table 1 were taken from water tank of building which is located in Jangareddygudem. The water quality parameters pH values, Turbidity values, temperature values are stored in SD card. Results of the water in water tank is slightly neutral i.e., pure water and having very less turbidity.

Table 1. Water Quality Measurement-1 at Water tank, Jangareddygudem Dated 23-05-2022

pH values	Turbidity values (NTU)	Temperature (°C)	Latitude	Longitude
6.73	4.45	34.44	17.1297595	81.3017427
7.01	4.63	34.63	17.1297400	81.3017377
6.83	4.65	34.75	17.1297424	81.3017381
6.74	4.63	34.50	17.1297560	81.3017337
6.67	4.69	34.38	17.1297618	81.3017333
6.62	4.66	34.25	17.1297638	81.3017353
6.74	4.67	34.25	17.1297554	81.3017276
6.87	4.68	34.44	17.1297560	81.3017337
6.83	4.66	34.31	17.1297583	81.3017358
7.02	4.29	34.13	17.1297695	81.3017447
7.03	4.32	34.00	17.1297570	81.3017357
6.93	4.21	34.02	17.1297485	81.3017390

Table 2. Water Quality Measurement-2 at Lake view, Vishnu Institute of Technology, Sri Vishnu Educational Society, Bhimavaram, Dated 24-05-2022

pH values	Turbidity values (NTU)	Temperature (°C)	Latitude	Longitude
7.23	4.65	39.13	16.5663129	81.5215628
7.08	4.63	39.19	16.5663167	81.5215554
6.97	4.69	39.31	16.5663309	81.5215413
6.82	4.66	39.50	16.5663167	81.5215467
6.78	4.67	39.56	16.5663293	81.5215390
6.70	4.67	39.25	16.5663492	81.5215185
6.67	4.68	39.90	16.5663726	81.5215306
6.64	4.66	39.85	16.5663717	81.5215437
6.65	4.29	39.75	16.5663646	81.5215531
6.63	3.68	39.82	16.5663591	81.5215702
6.57	3.61	39.70	16.5663517	81.5215631

(continued)

Table 2. (*continued*)

pH values	Turbidity values (NTU)	Temperature (°C)	Latitude	Longitude
6.58	3.54	39.55	16.5663437	81.5215567
6.51	3.61	39.13	16.5663627	81.5215360
6.87	2.39	39.00	16.5663521	81.5215353
6.85	3.53	39.02	16.5663665	81.5215685

Table 2 shows the water quality monitoring results taken from a lake which is located near Vishnu Institute of Technology, Sri Vishnu Educational Society, Bhimavaram. The water quality parameters pH values, Turbidity values, temperature values are stored in SD card. Results of the water in the lake is slightly neutral and having turbidity.

Different conditions and stages were displayed on LCD in the system as shown in Fig. 5. Initially when water quality monitoring system is switched on LCD shows display 1, then it checks SD card is connected or not, if SD card is connected, we get display 2 indication else we get display 3 indication. Display shows the sensor values as shown in Display 4. If system is sending message alert to user display as shown in Display 5, when message alert was sent then the display will be as shown in Display 6.

LED indications are shown in Fig. 6. When water quality monitoring system is switched on then power indication LED will be on. If SD card is storing the data, we get an LED blinking indication. Similarly, When GPS module is getting data of coordinates then we get LED blinking indication.

SMS messages shown in Fig. 7 are the alert messages to the user. When the device is switched on, then the user will get a message as "GSM based Water Quality Monitoring RC boat is started". If the values were abnormal when compared with predefined values then user get an alert message with sensor values and location as "pH value = ..., Turbidity value = ..., Temperature value = ..., "Pollution found in water in this location","https://www.google.com/maps/?q=17.120454,81.298873". If RC boat is about to sink when it is detected by water sensor, user get an message alert along with location as "RC Boat is about to sink, Rescue the boat", https://www.google.com/maps/?q=17.120454,81.298873".

Display 1

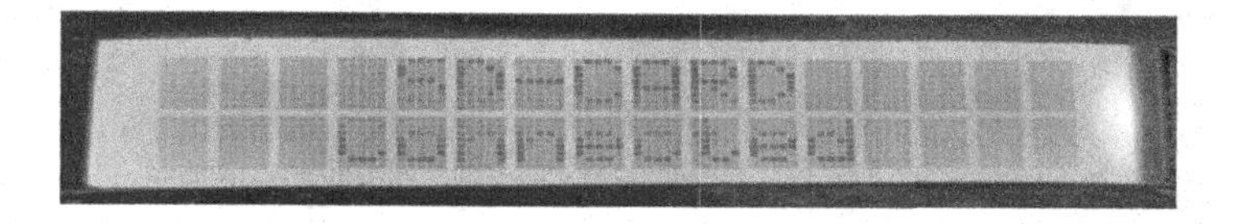

Display 2

Display 3

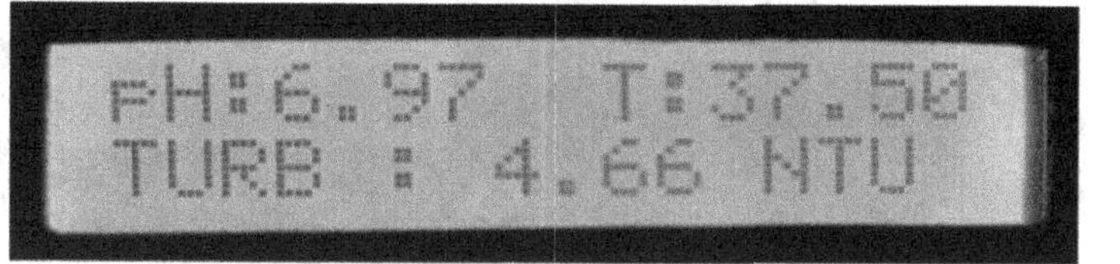

Display 4

Display 5

Display 6

Fig. 5. LCD display at different stages

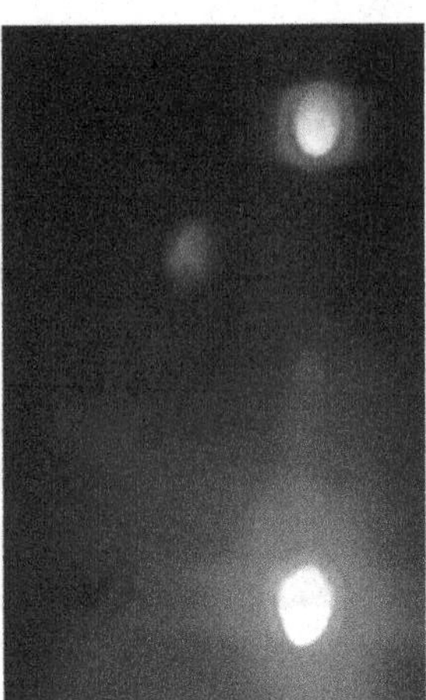

Fig. 6. LED indications

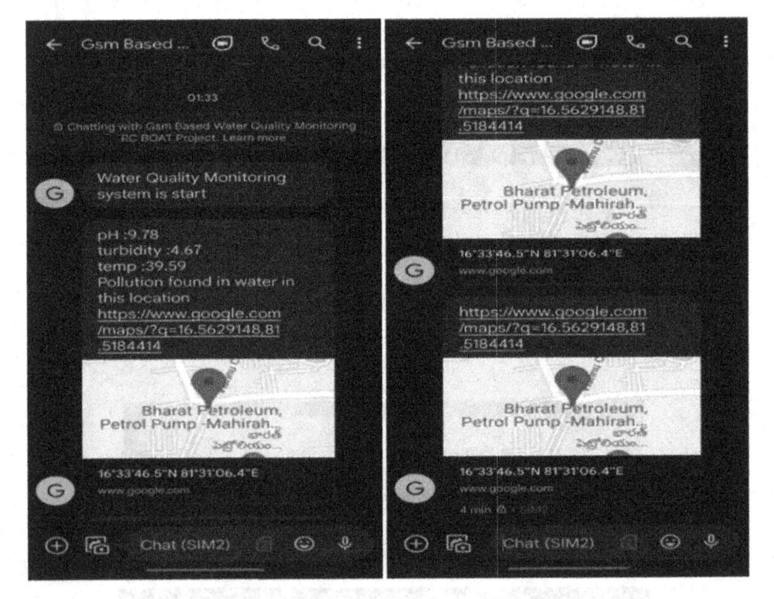

Fig. 7. SMS alerts-1 and 2

5 Discussions

Water is such a valuable resource for the existence of many living beings. Monitoring of water quality is very crucial for the healthy existence of living beings. Due to the scarce water resources, increasing population, and aging infrastructure, it is becoming a challenging task to provide good quality water to all living beings. As a result, better strategies for observing of state of water and novel methods of water characterization are required. Although current approaches analyze physical, chemical, and biological agents, they have significant flaws, including inadequate spatiotemporal coverage, labor-intensive and high costs (people, operation, and equipment). Requirement of a system to provide real-time water quality data to enable public health choices is very essential in the present scenario. As a result, constant water quality monitoring is required. The proposed system is a low cost, real time water quality monitoring system consisting of different hardware modules such as Remote Control Module, Remote Control Boat, and a water quality monitoring system embedded with various sensors. Using this system, water quality was monitored at various locations having difference in quality of water. From the measurements, it was observed that the values of PH, temperature and Turbidity of water were varied depending on the quality of water indicating the whether the water is suitable for drinking or not. This kind of real time water monitoring assists the local authorities, Government authorities to understand the water quality and make the required arrangements to process it for better quality before supplying the water to the public. The proposed system can be modified for constant monitoring of water quality. Impure quality of water can be tested in laboratories for further analysis. Further the system can be modified to monitor the level of water present in lakes and reservoirs which are the main sources of drinking water. "Internet of Things" can be used to let everyone acknowledge about the water quality.

6 Conclusions

The proposed system is used to monitor the quality of water remotely, get the real time data and also this system alerts the remote user if the water quality is abnormal. The proposed system is used to test the quality of water in different locations and water bodies such as local water tanks, lakes, ponds, etc. Measurements were taken using the proposed system and observed the PH, temperature and turbidity of various water bodies. Variations in PH, temperature, and turbidity values were observed for different water bodies. These measurements enables the remote user to understand and analyze different water bodies and see whether the water is acidic, basic in nature, its turbidity is good or not for the health of living beings. This kind of real time water monitoring assists the local authorities, Government authorities to understand the water quality and make the required arrangements to process it for better quality before supplying the water to the public.

References

1. Kale, V.S.: Consequence of temperature, pH, turbidity and dissolved oxygen water quality parameters. Int. Adv. Res. J. Sci. Eng. Technol. **3**(8), 186–190 (2016)
2. Adu, B.W., Oyeniyi, E.A.: Water quality parameters and aquatic insect diversity in Aahoo stream, southwestern Nigeria. J. Basic Appl. Zool. **80**(1), 1–9 (2019)
3. Amrita, C.M., Babiyola, D.: Analysing the water quality parameters from traditional to modern methods in aquaculture. Int. J. Sci. Environ. Technol. **7**(6), 1954–1961 (2018)
4. Zhu, B., et al.: Water quality impacts of small-scale hydromodification in an urban stream in Connecticut, USA. Ecol. Process. **7**(1), 1–10 (2018). https://doi.org/10.1186/s13717-018-0122-z

5. Toivanen, T., Koponen, S., Kotovirta, V., Molinier, M., Chengyuan, P.: Water quality analysis using an inexpensive device and a mobile phone. Environ. Syst. Res. **2**(1), 1–6 (2013). https://doi.org/10.1186/2193-2697-2-9
6. Simbeye, D.S., Yang, S.F.: Water quality monitoring and control for aquaculture based on wireless sensor networks. J. Netw. **9**(4), 840 (2014)
7. Demetillo, A.T., Taboada, E.B.: Real-time water quality monitoring for small aquatic area using unmanned surface vehicle. Eng. Technol. Appl. Sci. Res. **9**(2), 3959–3964 (2019)
8. Koparan, C., Koc, A.B., Privette, C.V., Sawyer, C.B.: In situ water quality measurements using an unmanned aerial vehicle (UAV) system. Water **10**(3), 264 (2018)
9. Mei, K., Zhu, Y., Liao, L., Dahlgren, R., Shang, X., Zhang, M.: Optimizing water quality monitoring networks using continuous longitu dinal monitoring data: a case study of Wen-Rui Tang River, Wenzhou, China. J. Environ. Monit. **13**(10), 2755–2762 (2011)
10. Demetillo, A.T., Japitana, M.V., Taboada, E.B.: A system for monitoring water quality in a large aquatic area using wireless sensor network technology. Sustain. Environ. Res. **29**(1), 1–9 (2019)

Smart City Eco-System and Communications

Study of Smart City Compatible Monolithic Quantum Well Photodetector

Prakash Pareek[1](✉), Naveen Kumar Maurya[1], Lokendra Singh[2], Nishu Gupta[3], and Manuel J. Cabral S. Reis[4]

[1] Department of Electronics and Communication Engineering, Vishnu Institute of Technology, Bhimavaram 534202, Andhra Pradesh, India
prakash.p@vishnu.edu.in

[2] Department of ECE, Koneru Lakshmaiah Education Foundation, Vaddeswaram 522002, Andhra Pradesh, India

[3] Department of Electronic System, Faculty of Information Technology and Electrical Engineering, Norwegian University of Science and Technology, 2815 Gjøvik, Norway

[4] Engineering Department, UTAD/IEETA, 5001-801 Vila Real, Portugal

Abstract. This work is focused on the exploration of the potential of Group IV alloy based nanostructured photodetector which can perform well in smart city environment. The theoretical model for multiple quantum well detector based on Germanium Tin (GeSn) alloy working in short wave infrared range (SWIR) is proposed after considering various design aspects. In this work, detectivity is estimated for SiGeSn/GeSn based interband multiple quantum well infrared photodetector. Detectivity is calculated by using responsivity and dark current, assuming appropriate conditions. After calculation, it is studied under variation of some important parameters. The result reveals the enhancement of detectivity with number of wells. It also indicates that the low biasing range is quite sufficient to make the proposed device work efficiently. Moreover, peak detectivity in the tune of 10^9 cm $Hz^{1/2}$ W^{-1} is attained.

Keywords: Smart City · GeSn · Detectivity · Group IV · MQWIP · Interband

1 Introduction

The concept of smart city aimed to advance the standard of human lives directly or indirectly. Smart cities provide all required infrastructure facilities and services more effectively than the conventional cities. Recently, with the advent of internet of things (IoT), the doors are opened for the city administrators to make the real implementation of smart city smoothly. Smart city really helps to incorporate advance and state of art technologies in various dimensions of city life like education, healthcare, sanitation, water supply, commutation services, among other. IoT proved to be a boon for providing smart city environment with the feasibility of automation in crucial aspects of citizen life [1]. However, it also manifests a large amount of data required in the tune of Zetta Bytes (ZB) or more. Unfortunately, it is just the beginning, because the demand of data is

N. Gupta et al. (Eds.): IC4S 2022, LNICST 472, pp. 215–224, 2023.
https://doi.org/10.1007/978-3-031-28975-0_18

going to increase at an alarming rate. As a result of this, surge in energy requirement will emerge. Hence, data centers are expected to consume a big chunk of the world electric power.

Considering the above two points, requirement of high-speed processing as well as low energy consumption technology is first in the bucket list of any smart city technical administrator. Microelectronics engineers are working superbly to design ultra-low energy consumption and high data processing integrated chips which can consume up to pico-Joule of energy. Having said that, it is not sufficient energy saving which is expected in the smart city environment because of the large number of sensors and actuators required for signal processing. Concurrently, photonics also showed a tremendous surge specially after the invention of Laser. There is no doubt regarding the speed of this technology due to mass less photons which carry the data. Another important benefit of photonics in smart city is the low energy consumption of the photonic integrated circuit which is in the range of femto-Joule.

Due to the aforementioned advantages, photonic integrated circuit research field witnessed a large amount work done by the researchers in recent times [2]. In this integrated circuit, a single platform or substrate accommodate all components like detector, source and many more. As a result, we can design a low cost CMOS compatible monolithic integrated device [3]. Moreover, data is also transmitted as well as received at higher rate due to all optical processing in this monolithic integrated device which seems to be perfect in the smart city eco system. One of the indispensable parts of this device is a competent integrated detector, which is responsible for receiving and sensing the optical signal and convert it into its electrical counterpart.

Within this frame of reference, research on some important aspects of integrated photodetector and its fabrication is very much relevant. But before its fabrication, a detailed theoretical study for its suitable design should be done to validate its feasibility. The conventional material from III-V group family is not suitable due to difficulties in their growth on Silicon platform [4]. Moreover, the material processing and fabrication cost also increase for III-V materials like GaAs, InGaAs, among other. As a result, workers have to look for another option for an appropriate material which can fulfill the above criteria to serve as the material for integrated detector.

Considering the above points, the focus of the researchers shifted towards Silicon based materials like Si, Ge, etc. The biggest bottleneck for the realization of monolithic photodetector is the indirect bandgap nature of Si, Ge. But one property of Ge had given hope to researchers for the realization of all group IV photodetector. In Germanium (Ge), there is very little gap between the indirect bandgap and direct band gap. In other words, Γ and L band edges in Germanium are separated by only few mV [5]. This characteristic had provided the opportunity to make the direct bandgap edge lower than the indirect bandgap edge so that Ge act like a direct bandgap material. There are different types of techniques of doing that like 'n' doping in Ge, introducing strain in Ge and incorporation of Sn into Ge [6]. Among these methods, the third one is standout and received very well by the researchers. The findings of the wide-ranging work related to Sn incorporation in Ge are very interesting and have paved the path of the possible role of GeSn/SiGeSn alloy in integrated photodetector [7, 8].

But before the actual realization of SiGeS/GeSn based sensors, one need to assess their validity theoretically. As a result, models need to be proposed which obey basic device physics and also practical fabrication aspects. The theoretical study is having its own limitations due to dearth of reported works in SiGeSn/GeSn photodetector regime [8]. In addition, properties of Sn and Ge make this task more challenging. For instance, lattice constant difference between Ge and Sn is very large. Excessive mismatching cause strain gross imperfection during fabrication if ignored. This makes design of theoretical model for SiGeSn/GeSn detector more challenging.

In order to resolve this issue, a theoretical model for GeSn/SiGeSn Quantum well infrared photodetector (QWIP) was introduced by one of the authors [9, 10]. In this model, the strain factor was compensated by introducing some modifications which are explained later. Different aspects of QWIP performance were investigated in single well and multiple well configuration [9, 10]. Having said that, minimum detectable signal still needs to be studied under different variables for GeSn/SiGeSn QWIP. Detectivity is a most important parameter from a device designer perspective and proper effort should be devoted towards its exploration. Because in case of all group IV based QW based detector, detectivity is hardly reported.

Thus, in this work, photosensitive device compatible to smart city applications is studied along with its crucial parameters. This device is analyzed by introducing a theoretical model of group IV based MQWIP. The remaining sections of this paper will give a brief summary about the studied model, methodology and a detailed discussion on the results obtained.

2 Design of MQWIP

The theoretical model for considered MQWIP is provided in Fig. 1. It possesses multiple quantum well structure. Each quantum well is formed by periodic repetition of single quantum well (SQW) which consists of GeSn active well which is surrounded by SiGeSn barriers on either side. The bandgap of SiGeSn is higher than GeSn active well, which is a type I quantum well configuration [11]. Moreover, the thickness of the well is carefully chosen to allow single state in each of conduction band and valence band. The mole fraction of Sn in GeSn well is selected so that GeSn become direct bandgap in nature [9]. The device operational wavelength as per design is 3.4 μm.

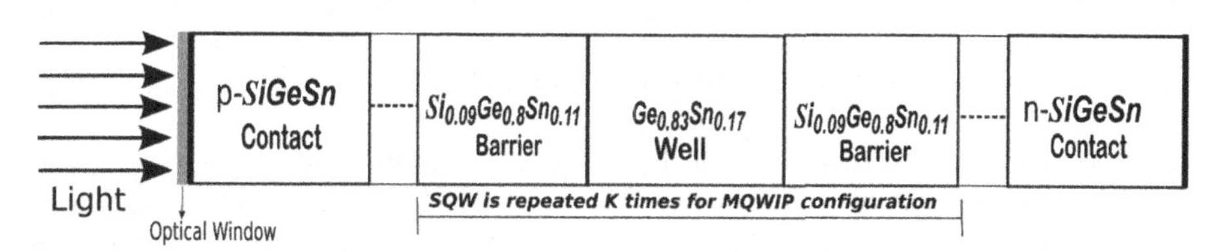

Fig. 1. Considered multiple quantum well infrared photodetector model

In the design of the considered QWIP structure, each quantum well structure consist of one GeSn narrow bandgap well and two SiGeSn wide bandgap barriers. This structure is repeated 'K' times in a periodical manner. Now regarding the content of well, 17% of Sn is considered which is sufficient to induce direct band gap nature in the GeSn layer [10]. The SiGeSn layer acts as a barrier with 9%, 80% and 11% of Si, Ge and Sn content respectively. The width of the barriers is also one of the important aspects in design of this QWIP. This is because of the involvement of alloys like GeSn and SiGeSn which are difficult to grow in normal conditions. The large lattice mismatching between Sn and Ge/Si tempts the researchers to look for other ways to successfully fabricate the GeSn and SiGeSn layers. Dislocation can occur if this problem cannot be handled properly. Hence in order to avoid this issue, the strain compensating technique is used, which is already successfully applied in case of GeSn based quantum well lasers [12, 13]. The dimension of barriers is chosen according to following strain balance environment equation as below [13],

$$\sum_{i=1}^{p} \frac{X_i j_i e_i}{l_i}; \; X_i = E_{11}^{(i)} + E_{12}^{(i)} - 2\frac{E_{12}^{(i)2}}{E_{11}^{(i)}} \tag{1}$$

where p is the number of layers, the coefficients of elasticity are E_{11}, E_{12}, the width of the layer is represented by j, and e denotes the dielectric constant. In this strain balancing environment, well and barrier experience opposite strain. In our design, well is chosen to be under compressive strain whereas barriers are under tensile strain. By using the above expression (Eq. 1), the width of the barrier is attained as 36 Å. A buffer layer is also considered to make fabrication feasibility and strain compensating conditions favorable. In Fig. 1, P and N contact layers are also shown. The compositions of these layers are chosen to make them lattice matched to buffer. The input electromagnetic signal (light) is considered to be propagate in transverse electric mode owing to the compressive strain in well [11].

3 Performance Evaluation

Sensitivity of the sensor is a very important decisive parameter in its performance. In terminology of photonic detection, it is termed as detectivity. It is almost mandatory to specify detectivity for device engineers as well as manufacturers. Actually, detectivity indicates the minimum signal which can be detect by the sensor without any noise. The detectivity, DMQWIP can be calculated with the help of Eq. (2) [4]:

$$D_{MQWIP} = \frac{1}{2}(R_{espK})\left[q\left(Id_{darkK\_electrons} + Id_{darkK\_holes}\right)\right]^{-1/2} \tag{2}$$

where R_{epsK} indicates responsivity for 'K' quantum well periodic structure (as shown in Fig. 1), the current density in absence of light denoted by $Id_{darkK\_electrons}$ and Id_{darkK_holes} respectively. As it is clear from the above equation, evaluation of D_{MQWIP} requires the values of current in presence of light (in the form of responsivity) and absence of light (dark current).

Current is generated by the movement of charge carriers triggered by the incident light. this current can be calculated in terms of responsivity. In order to obtain this parameter, the charge mechanism of MQWIP should be studied in detail. Further, the use of the continuity equation becomes crucial here. In MQWIP, the charge carriers are moving across multiple interfaces and also there is an interaction between carriers of adjacent well in multiple quantum well structure. The light absorption parameter is also playing a vital role in this analysis.

After responsivity, the sensitivity calculation requires the current which is generated in the absence of light. The calculation of this current is not straightforward, and it requires certain assumptions and considering special conditions which are as follows. Firstly, in the absence of light, no charge carriers are generated, that is why the corresponding parameter should be neglected in the rate equation. Now if light is not present to trigger the charge carrier movement, temperature role become very important. So, instead of optical rate, thermal rate of generation is considered in the calculation of the current in absence of light. As we are considering interband transition in this case, charge carriers relaxation time in interband state also plays a key role in this calculation. The generation rate of carriers, g_d, in dark conditions can be calculated as [11]

$$g_d = \frac{n_{int,2D}(T)}{t_{interband\_relaxation}} \tag{3}$$

In the above expression $n_{int,2D}$ is the temperature dependent carrier density in intrinsic state and $t_{interband\_relaxation}$ is the relaxation time of carriers in interband state.

Now, current in absence of light both for holes and electrons can be obtained by using the rate equation under DC conditions. Some changes are required to make in the conventional rate equation and it is given as [14],

$$\frac{\delta n_{QW,dark}}{\delta t} = \frac{cap_p}{q} id_{max,e} + g_d - n_{QW,dark} R_{total} \tag{4}$$

In the above expression, the generated carriers in dark conditions are denoted by $n_{QW,dark}$, the peak value of current density in absence of light indicated by $id_{max,e}$ for a solo well. The combination of carrier rates by different mechanism like optical generation, thermionic and tunneling are shown by R_{total}. Moreover, in this calculation, the model for MQWIP coined by Ryzhii is adopted, hence the probability of capture of carriers is used here, which is denoted by cap_p. This calculation is implemented for multiple quantum well structure and considering the process reported by Ryzhii [15]. Then after getting the values of responsivity and dark current density, sensitivity can be calculated for proposed MQWIP as shown in Eq. 2.

Another important parameter for study is the distribution of the electric field, which is most crucial when the device is to be designed for smart city environment. The calculation of the electric field and its spatial distribution in a quantum well is evaluated with the help of the Ryzhii model [16]. The emitter layer and tunneling layer are considered in this model. The current density of emitter layer, j_e, is related to the electric field of the tunneling layer, E_t, as shown in Eq. 5.

$$\frac{j_m}{j_e} = \exp\left(\frac{E_t}{E + ME_D\left(1 - \frac{n_{QW}}{n_{i,2D}}\right)}\right) \tag{5}$$

where j_m is the maximum emitter current density, E is the electric field over the active sensing region, n_{QW} is the electron concentration in quantum well, and E_D is the field due to intrinsic sheet concentration $n_{i,2D}$.

On further simplifying Eq. 5,

$$\left(1 - \frac{n_{QW}}{n_{i,2D}}\right) = \left(\frac{E_t}{\ln(j_k)} - E\right)\frac{1}{ME_D} \tag{6}$$

where $j_k = (j_m/j_e)$, and j_e can also be expressed in terms of tunneling electric field in another way as [16],

$$j_e = j_m \exp\left(\frac{-E_t}{E_e}\right); E_e = \frac{E_t}{\ln(j_k)} \tag{7}$$

Now, position dependent potential can be written by using the Poisson equation for quantum well and it is written as [16],

$$p_d(x) = V\frac{x}{W} + \frac{2\pi q n_{i,2D}}{\varepsilon w_p}.x(W - x)\left(1 - \frac{n_{QW}}{n_{i,2D}}\right) \tag{8}$$

where V is the applied bias, p_d is the position dependent potential, w_p is the quantum well period ($w_d + w_B$), ε is the dielectric constant, x is the position coordinate, and W is the total width of well. Then, the electric field for the first quantum well period is written as considering Fig. 1 and Ryzhii model [16],

$$E_1 = \left.\frac{d(p_d)}{dx}\right|_{x=w_p} = \frac{V}{W} + \frac{2\pi q n_{i,2D}}{\varepsilon w_p}.(W - 2w_p)\left(1 - \frac{n_{QW}}{n_{i,2D}}\right) \tag{9}$$

MQWIP is the main focused device structure. So, for the Mth quantum well, expression of the electric field is given by:

$$E_M = \left.\frac{d(p_d)}{dx}\right|_{x=Mw_p} = E + \frac{2\pi q n_{i,2D}}{\varepsilon w_p}(W - 2Mw_p)\left(1 - \frac{n_{QW}}{n_{i,2D}}\right) \tag{10}$$

From Eqs. 6, 7 and 10, the final expression for the electric field over MQWIP is written as:

$$E_M = \frac{E_t}{\ln(j_k)} + \frac{2\pi q n_{i,2D}}{\varepsilon w_p}\left(\frac{E_t}{\ln(j_k)} - E\right)(W - 2Mw_p)\frac{1}{ME_D} \tag{11}$$

The above expression will be used to determine the distribution of the electric field across different wells of MQWIP.

4 Results and Discussions

As explained in the previous section, detectivity is evaluated by calculating the photo current and the current in absence of light. The value of the photo current, taking into consideration of carrier mechanism, had already been published by the first author for single quantum well detector [15]. In this work, a similar procedure was adopted to calculate the photo current for multiple quantum well photodetector. The current in absence of light is also obtained by adopting the process as explained in the previous section. Consider the number of well in the proposed model of MQWIP as 'M', then total current in absence of light can be shown under variation of 'M'. Therefore, in Fig. 2, Id_{dark} (summation of holes and electron current without light) is shown for different numbers of quantum well. It is clear from this figure that the current in absence of light reduces with enhancement in 'M'.

The carrier movement in absence of light is due to thermal effect. Thus, thermionic emission is the primary source of carrier generation, when there is no light. The trend in Fig. 2 is due to the following reason. When the number of well enhances, the effective field reduces. As the electric field drive the thermionic emission, this emission rate decreases on increasing the number of well. Reduction of the thermionic emission rate also cause decreasing of the current in absence of light. It can be also observed from the figure that after a particular value of QW number 'M', the dark current saturates, which was also reported by Ryzhii [16].

Equation 2 is used to calculate sensitivity or detectivity. It actually indicates the minimum amount of signal which can be detect by the device without any ambiguity. DMQWIP is calculated as explained in the previous section and the corresponding peak values are shown w.r.t. bias (V), for various 'M' in Fig. 3. It is revealed from the figure that as V increase, detectivity decreases, because the impact of V is more on the dark current rather than on responsivity. At minimum values of V, a maximum value of DMQWIP is obtained, i.e. 2.3 x 10^9cmHz$^{1/2}$W^{-1} for M = 14. Moreover, Fig. 3 also discloses that detectivity enhances with M. Another important point depicted from this figure is that low biasing is sufficient for good sensitivity for the proposed detector. In smart city environment, low range of biasing will be ideal as energy consumption would be low, which is desirable.

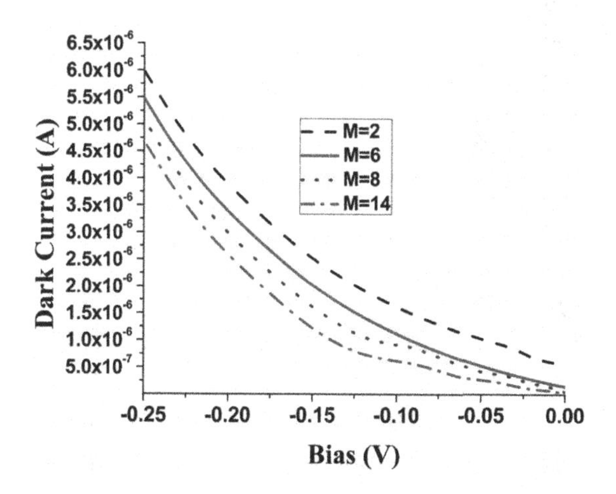

Fig. 2. Current without light input for various M

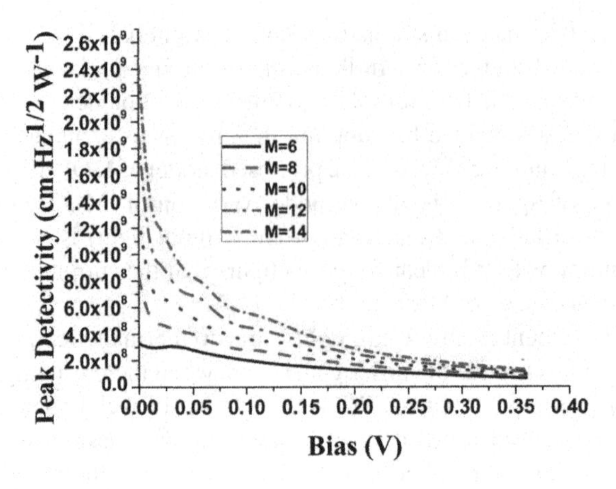

Fig. 3. D_{MQWIP} for various M

Now, the spatial distribution of the electric field for M = 8 is calculated and shown in Fig. 4 at applied bias 0.1 V. It can be observed from the figure that the minimum field in device, for M = 8, is in the saturation region. So, using the saturation velocity for our analysis seems to be appropriate. It can also be inferred from the figure that the first quantum well field is very high due to direct injection of emitter tunneling current into this well. Low biasing is good enough to operate this photosensitive device which can generate good amount of electric field.

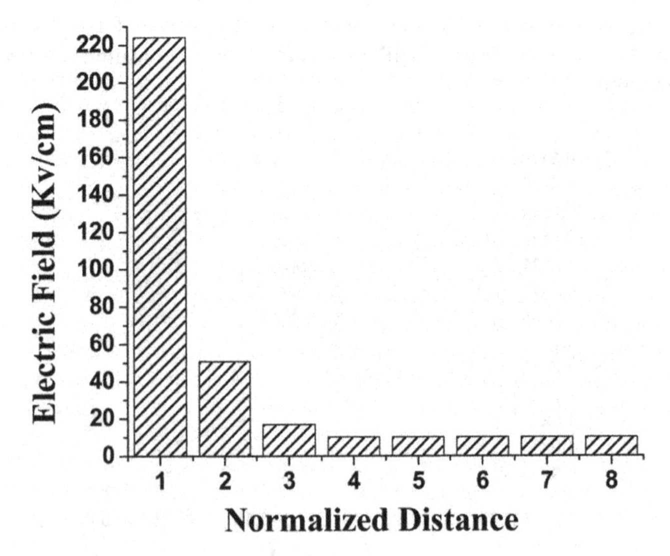

Fig. 4. Electric field distribution in MQWIP for M = 8

5 Conclusion

The present work focuses on the investigation of the potential of group IV based MQWIP for its compatibilities with smart city technologies by proposing a theoretical model. Sensitivity is calculated by considering various charge carrier mechanisms observed by multiple quantum well structure and its interfaces. The value of detectivity ($\sim 10^9$ cm $Hz^{1/2}W^{-1}$) obtained in this work is in the range of that of III-V based photodetectors, which is a very encouraging sign. It further motivates the researchers to explore more this device to make it suitable for one of the components of photonic integrated circuit monolithic chip. The outcome of this work depicts that the photonics is at the pole position among other emerging technologies to become an integral part of smart cities network. The proposed detector has immense potential to provide a solution of low energy consumption in smart city environments.

References

1. Da Silva, I.N., Flauzino, R.A.: Smart Cities Technologies, 1st edn. Intechopen, London (2016). https://doi.org/10.5772/61375
2. Ang, K., et al.: Silicon photonics technologies for monolithic electronic-photonic integrated circuit (EPIC) applications: current progress and future outlook. In: Proceedings of International Electron Devices Meeting (IEDM), pp. 1–4. IEEE, Baltimore (2009)
3. Roelkens, G., et al.: Silicon-based photonic integration beyond the telecommunication wavelength range. IEEE J. Sel. Top. Q. Electron. **20**(4), 394–404 (2014)
4. Bhattacharya, P.: Semiconductor Optoelectronic Devices, 2nd edn. Pearson Education Inc., New Jersey (1994)
5. Sze, S.M., Ng, K.K.: Physics of Semiconductor Devices, 3rd edn. Wiley-Interscience, New-Jersey (1969)
6. Goodman, C.H.L.: Direct-gap group IV semiconductors based on tin. In: IEEE Proceedings of the I: Solid State Electron Devices, vol. 129, pp. 189–192 (1982)
7. Kouvetakis, J., Menendez, J., Chizmeshya, A.V.G.: Tin based group IV semiconductors: new platforms for opto and microelectronics and silicon. Annu. Rev. Mater. Res. **36**, 497–554 (2006)
8. Soref, R.: Emerging SiGeSn integrated-photonics technology. In: Proceedings of IEEE Photonics Society Summer Topical Meeting Series, pp.100–101. IEEE, Newport Beach (2016)
9. Downs, C., Vandervelde, T.E: Progress in infrared photodetectors since 2000. Sensors **13**, 5054–5098 (2013)
10. Pareek, P., Das, M.K.: Theoretical analysis of direct transition in SiGeSn/GeSn strain balanced QWIP. Opt. Quant. Electron. **48**, 228 (2016). https://doi.org/10.1007/s11082-016-0498-x
11. Pareek, P., Das, M.K., Kumar, S.: Theoretical analysis of tin incorporated group IV alloy based QWIP. Superlattices Microstruct. **107**, 56–68 (2017)
12. Chuang, S.L.: Physics of Optoelectronic Devices, 2nd edn. John Wiley & Sons Inc., New York (1995)
13. Chang, C.E., Chang, S.W., Chuang, S.L.: Strain-balanced GezSn1-z-SixGey Sn1-x-y multiple-quantum-well lasers. IEEE J. Q. Electron. **46**, 1813–1820 (2010)
14. Ekin-Daukes, N.J., Kawaguchi, K., Zhang, J.: Strain-Balanced Criteria for Multiple Quantum Well Structures and Its Signature in X-ray Rocking Curves. Cryst. Growth Des. **2**(4), 287–292 (2002)

15. Pareek, P., Das, M.K., Kumar, S.: Responsivity calculation of group IV based inter band MQWIP. J. Comput. Electron. **17**, 319–328 (2018). https://doi.org/10.1007/s10825-017-1071-y
16. Ryzhii, V.: Characteristics of quantum well infrared photodetectors. J. App. Phy. **81**, 6442–6448 (1997)

Frequency Reconfigurable Antenna for 5G Applications at n77 and n78 Bands

Rangarao Orugu[1,2(✉)], Nesasudha Moses[1], and Doondi Kumar Janapala[1,2]

[1] Karunya Institute of Technology and Sciences, Coimbatore 641114, Tamilnadu, India
ranga.mouni@gmail.com
[2] Vishnu Institute of Technology, Bhimavaram 534202, Andhra Pradesh, India

Abstract. A frequency reconfigurable antenna is designed and presented for 5G applications. The proposed antenna is designed by using two T shaped patches positioned inversely to each other and connected by a PIN diode. Here the bottom T shaped patch is inverted and connected to the micro strip line feeding. FR4 substrate is used as dielectric and the backside ground is maintained partial with a rectangular open slot is etched opposite to the feed point. The PIN diode acts as a switch to change the operating frequency of the proposed antenna in-between n77 and n78 bands of 5G applications. The simulated analysis of proposed 5G antenna showed that, when diode is in OFF conditions it operates at 3.5 GHz with operational band covers from 3.34 GHz to 3.7 GHz. Similarly, when diode is in ON condition the antenna operated at 3.7 GHz with operational band covering from 3.47 GHz to 3.94 GHz. The performance of the frequency reconfigurable 5G antenna for OFF and ON conditions is compared with the help of S_{11} vs frequency, VSWR and radiation pattern curves and presented. The proposed antenna maintained similar radiation coverage at respective resonant frequencies for OFF and ON conditions making it suitable for future 5G applications.

Keywords: 5G · n78 band · n77 band · PIN diode · Frequency Reconfigurable

1 Introduction

The reconfigurable antennas capable of adjusting their performance based on the tuning provided can help in many ways in serving different applications by single antenna. An antenna which can operate at different band of frequencies when tuned can serve multipurpose. Similarly antenna having the capability of altering its radiation pattern to desired direction can help in many ways in communications systems in improving the signal strength in desired direction or reducing the coverage in unwanted region. These reconfigurable antennas garnered much attention in recent times. Frequency reconfigurable antenna where a single antenna can be made to serve different operational bands based on the switching condition provided has several advantages in multi application purpose communication systems. Different techniques are used to provide switching for reconfigurable antennas [1]. One of the most commonly used techniques is utilizing PIN diodes appropriate place in the antenna design [2].

N. Gupta et al. (Eds.): IC4S 2022, LNICST 472, pp. 225–233, 2023.
https://doi.org/10.1007/978-3-031-28975-0_19

Over the years different frequency reconfigurable antenna are being developed to serve various applications [3–8]. For UWB model the frequency selection can be based on which bands need to be passed or filtered, based on the requirement the operational bands can be tuned [9]. The frequency/radiation steerable antennas find applications majorly in Wireless Sensor Networks (WSN) [10]. In recent times reconfigurable antennas for LTE and 5G applications achieved more attention by researchers. These antennas are developed for lower frequency of operation to mm-wave applications [11–14]. In 5G new radio (NR), the n77 band covers from 3.3 GHz to 4.2 GHz with operational frequency at 3.7 GHz and n78 band covers from 3.3 GHz to 3.8 GHz with operating frequency maintained at 3.5 GHz [15, 16].

In the current work a single diode is used to switch the operating frequency of the antenna. The designed antenna has two T shaped patches connected by the diode and fed by micro-strip feeding. The diode OFF and ON conditions helped in reconfiguring the antenna operating frequency to operate at n78 and n77 bands respectively.

2 Frequency Reconfigurable 5G Antenna Design

2.1 5G Antenna Design

The proposed frequency reconfigurable 5G antenna design is illustrated in the following Fig. 1 and the geometrical dimensions in mm are listed in Table 1. The radiating element consists of two T shaped patches, where the first patch is positioned inversely. One side it is connected by a microstrip feed line and another side it is connected to the second T shaped patch via PIN diode. The PIN diode used in this design is BAR 64-02V diode. The diode forward and reverse biasing conditions are explained in the following Fig. 2. The diode OFF and ON conditions helped in switching the operating frequency from 3.5 GHz to 3.7 GHz. The ground plane of the antenna is maintained partial with an open slot at center part of the top side. The antenna used FR4 substrate with dielectric constant 4.4 and the overall dimensions of the antenna in XYZ directions are $25 \times 20 \times 1.6$ mm^3.

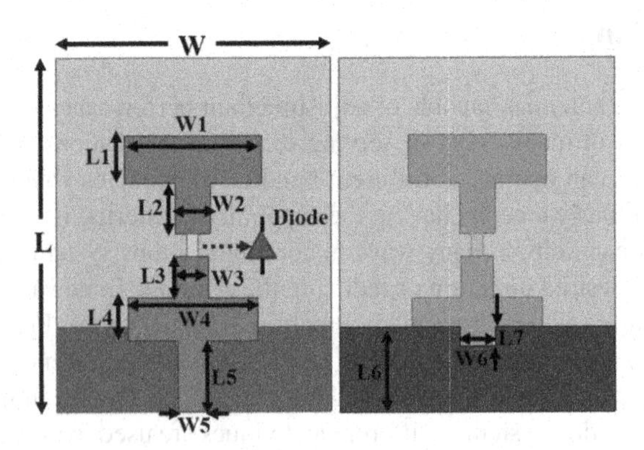

Fig. 1. Design of the proposed frequency reconfigurable 5G antenna

Table 1. 5G reconfigurable antenna design parameters in (mm).

Parameter	Value (mm)	Parameter	Value (mm)
L	25	W	20
L1	3.4	W1	10
L2	3.5	W2	2.6
L3	3	W3	2.4
L4	3	W4	9.5
L5	5	W5	2.2
L6	6	W6	2.6
L7	1.4		

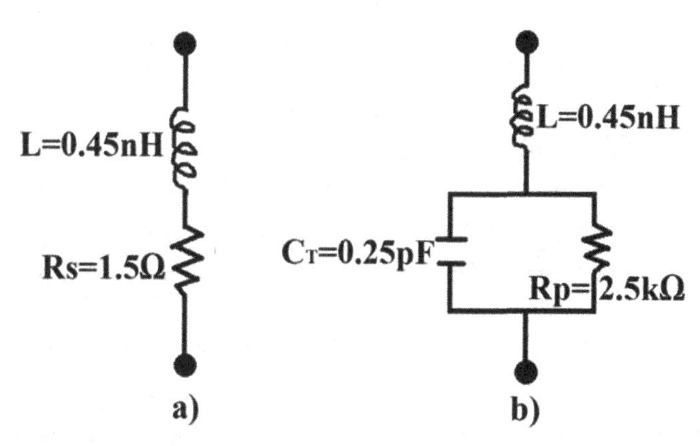

Fig. 2. BAR 64-02V PIN diode biasing conditions

3 Parametric Study

The above-mentioned suitable dimensions in Table 1 are obtained from parametric study. Change in some of these parameters can affect the performance of the antenna. The parametric study analysis for few selected dimensions with the help of S_{11} vs frequency curves are presented in this section.

3.1 W1, W2, W4, L3 and L4 Dimensions Parametric Study

The dimensions selected for parametric study are from the two T shaped patches. The top T patch dimension W1 and W2 effect on antenna performance for diode OFF and ON conditions are presented in the following Figs. 3 & 4. Where it can be observed from Fig. 3 that the dimension W1 effects the antenna operating frequency drastically as W1 increases, the operating frequency shifts towards lower resonance with a decline in impedance matching. Here for W1 = 10 mm the antenna maintained operating frequency near at 3.5 GHz for OFF condition and at 3.7 GHz for ON condition.

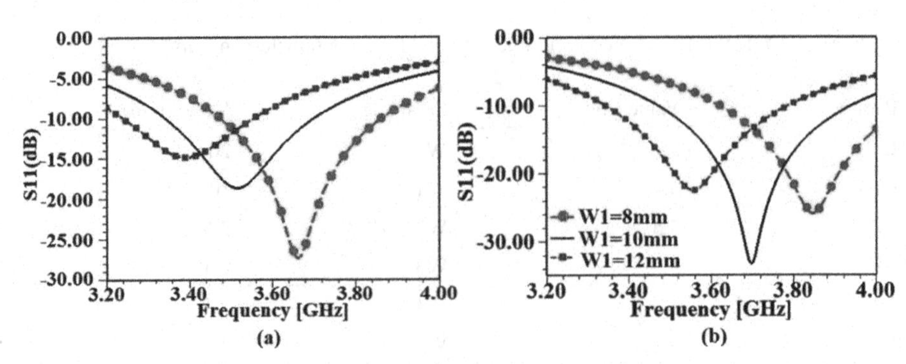

Fig. 3. S_{11} vs Frequency curve comparison for W1 variation a) Diode OFF and b) Diode ON

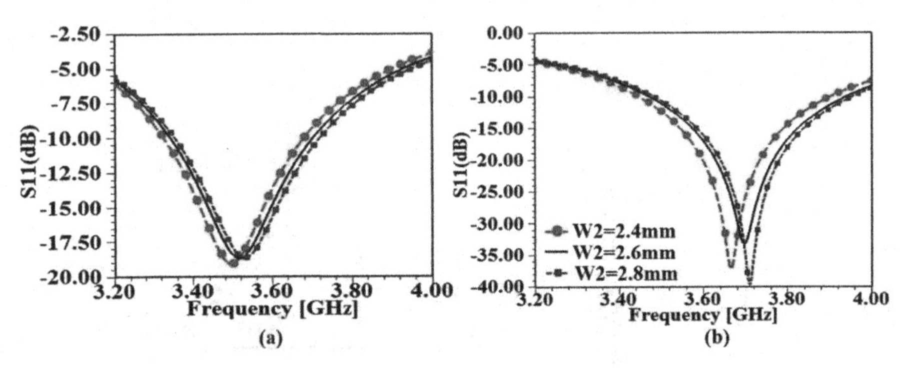

Fig. 4. S_{11} vs Frequency curve comparison for W2 variation a) Diode OFF and b) Diode ON

Similarly, from Fig. 4 for dimension W2 the shift in operating frequency is less as the step size of the dimension is less. In view of the required operating frequencies for n78 and n77 bands W2 = 2.6 mm is selected.

The bottom T patch dimensions W4, L3 and L4 varied and the S_{11} vs frequency curves comparison is presented in the following Figs. 5, 6 and 7.

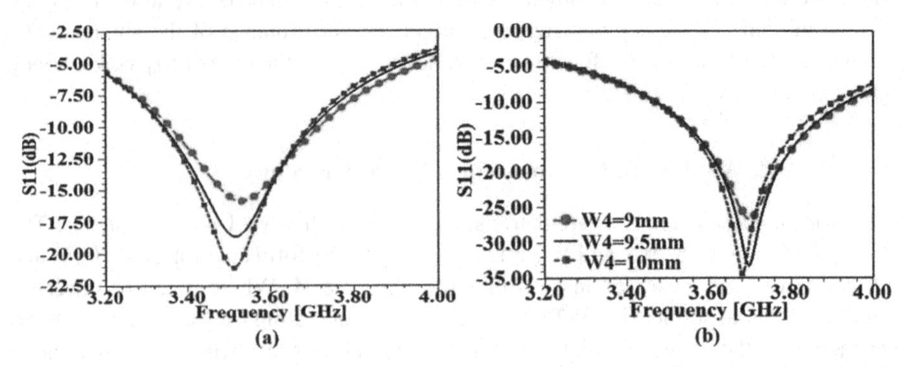

Fig. 5. S_{11} vs Frequency curve comparison for W4 variation a) Diode OFF and b) Diode ON

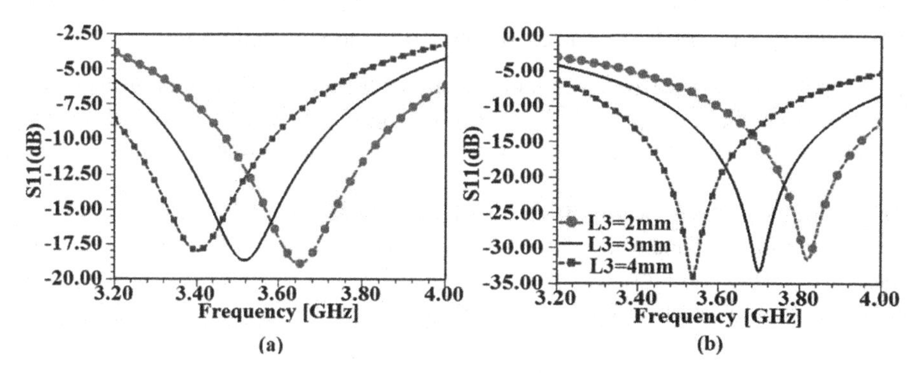

Fig. 6. S_{11} vs Frequency curve comparison for L3 variation a) Diode OFF and b) Diode ON

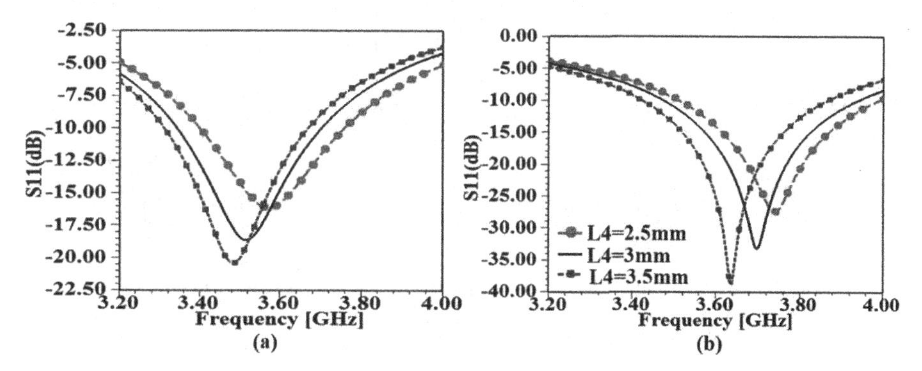

Fig. 7. S_{11} vs Frequency curve comparison for L4 variation a) Diode OFF and b) Diode ON

From Fig. 5 it can be seen that, for W4 variation effect on operating frequency shift is less for both OFF and ON conditions of the diode. But the impedance matching of the antenna is affected, the increment in W4 dimension helped in improving the impedance matching. Based on the required operating frequencies for the considered 5G bands W4 = 9.5 mm is considered for further analysis. For dimensions L3 and L4 of the operating frequency is being shifted drastically when these dimensions are varied. When L3 is increased the operating frequency is being shifted towards the lower resonance while the impedance matching is affected less. But for increment in dimension L4 not only the operating frequency is being shifted towards lower resonance, the impedance matching is improving. Based on the operating frequencies of n78 and n77 bands L3 = 3 mm and L4 = 3 mm are selected. Similarly rests of the dimensions in antenna geometry are varied and through parametric study the suitable dimensions that produced required operating frequencies are listed in Table [1].

4 Results and Discussions

The performance comparison for the PIN diode OFF and ON conditions are explained in this section using S_{11} vs frequency, VSWR, radiation patterns and surface current distribution field curves.

4.1 Reflection Coefficient Comparison for OFF and ON Conditions

The proposed antenna performance comparison for OFF and ON conditions using reflection coefficient curves is presented in the following Fig. 8. The designed 5G antenna operated at 3.5 GHz with operational band below −10 dB, covering from 3.34 GHz to 3.7 GHz (n78 band) for diode OFF condition. Similarly, it operated at 3.7 GHz with operational band covering from 3.47 GHz to 3.94 GHz (n77 band). So, the 5G antenna-maintained band widths of 360 MHz and 470 MHz for OFF and ON conditions respectively. The return loss for OFF condition at 3.5 GHz is maintained at −18.41 dB and for ON condition at 3.7 GHz the return loss is −33 dB.

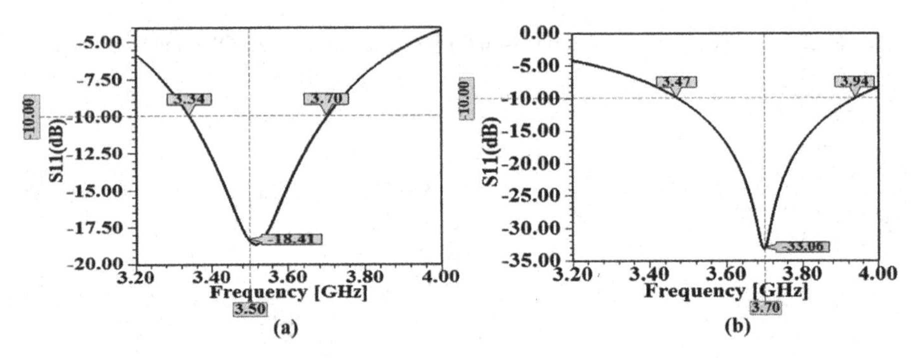

Fig. 8. S_{11} vs Frequency curves comparison a) Diode-OFF and b) Diode-ON

4.2 VSWR Comparison for OFF and ON Conditions

Similarly for diode OFF and ON conditions the VSWR curves comparison is presented in the following Fig. 9.

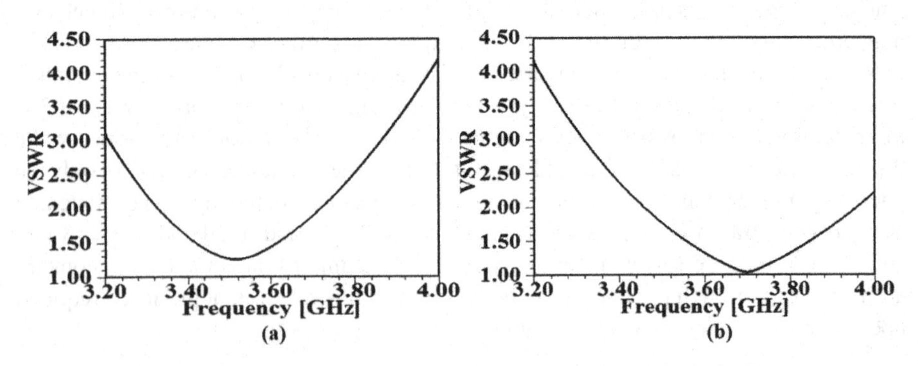

Fig. 9. VSWR curves comparison a) Diode-OFF and b) Diode-ON

From the above Fig. 9, it can be observed that the VSWR value for the proposed antenna maintained in between 1 and 2 values for the respective operational bands during OFF and ON conditions.

4.3 2D and 3D Radiation Patterns Comparison for OFF and ON Conditions

As the antenna resonant frequency maintained at 3.5 GHz for diode OFF condition and at 3.7 GHz for diode ON condition the radiation pattern comparison is given at the respective resonant frequencies in the below Fig. 10.

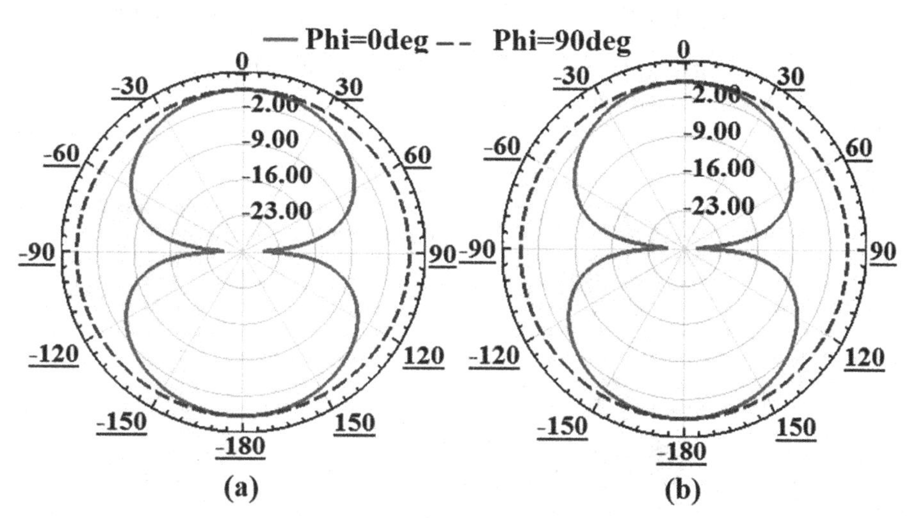

Fig. 10. 2D radiation pattern comparison a) At 3.5 GHz for Diode-OFF and b) At 3.7 GHz for Diode-ON

From Fig. 10 it can be seen that, the 5G antenna maintained similar radiation coverage for the diode OFF and ON conditions at its respective operating frequencies proving its suitability for 5G applications.

Similarly the 3D radiation patterns at 3.5 GHz and 3.7 GHz for OFF & ON conditions are compared in the following Fig. 11.

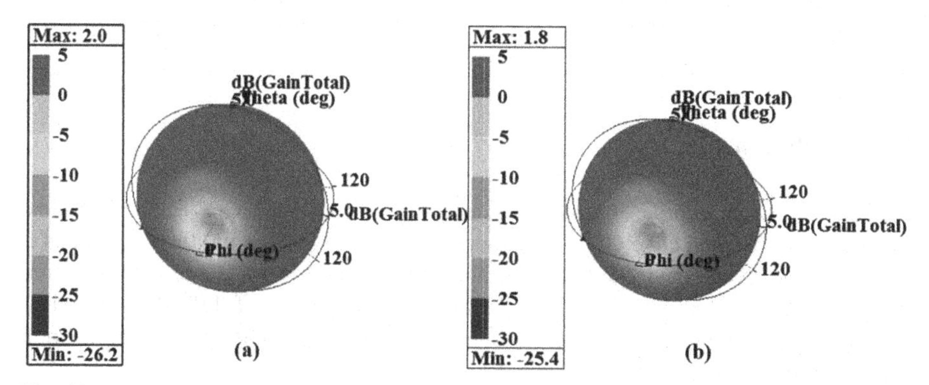

Fig. 11. 3D radiation pattern comparison a) At 3.5 GHz for Diode-OFF and b) At 3.6 GHz for Diode-ON

From the above Fig. 11 it can be observed that, the 5G antenna maintained similar shape for radiation coverage in overall. But the maximum gain during diode OFF condition is maintained at 2 dB at 3.5 GHz resonant frequency. Whereas for diode ON conditions the maximum gain is 1.8 dB at resonant frequency 3.7 GHz.

4.4 Surface Current Distribution Comparison for OFF and ON Conditions

The current distribution for respective resonant frequencies during diode OFF and ON conditions is presented in the following Fig. 12.

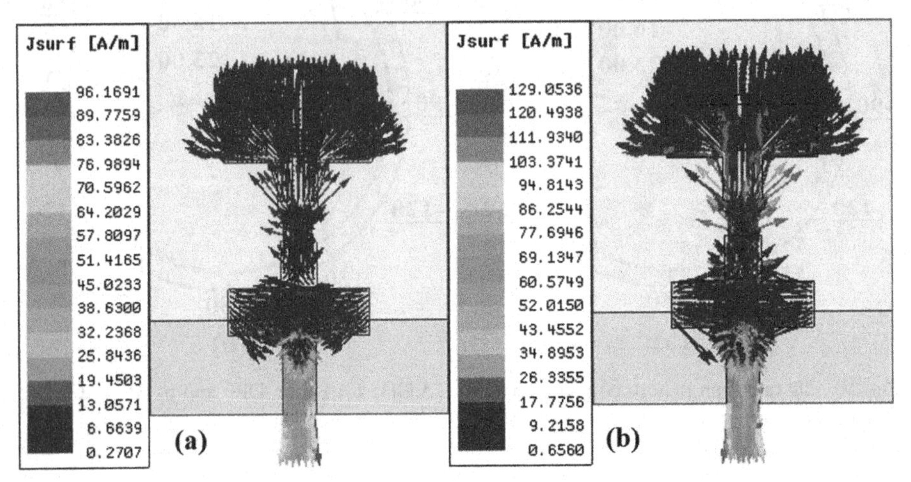

Fig. 12. Surface current distribution comparison a) At 3.5 GHz for Diode-OFF and b) At 3.7 GHz for Diode-ON

From the current distribution curves shown in above Fig. 11 it can be seen that at respective operating frequencies while the diode is OFF and ON the 5G antenna maintained different current distribution through the patch. For diode OFF condition the maximum J value at 3.5 GHz is 96.16 A/m and for ON condition it is 129.05 A/m at 3.7 GHz.

5 Conclusion

A frequency reconfigurable 5G antenna is designed and presented for n78 and n77 bands. The designed antenna covered n78 band operating at 3.5 GHz while the diode is OFF condition and maintained bandwidth of 360 MHz. Similarly for diode ON condition it serves 5G n77-band operating at 3.7 GHz with bandwidth of 470 MHz. Antenna exhibited good radiation coverage for both OFF and ON conditions making it good candidate for 5G applications.

References

1. Ojaroudi Parchin, N., Jahanbakhsh Basherlou, H., Al-Yasir, Y.I.A., Abdulkhaleq, A.M., Abd-Alhameed, R.A.: Reconfigurable antennas: switching techniques—a survey. Electronics **9**(2), 1–14 (2020)
2. Jayamani, K., Indhumathi, K., Jeevakumari, S.A.A.: A survey on frequency reconfigurable antennas using passive element for ISM band. In: 2020 IEEE 7th International Conference on Smart Structures and Systems (ICSSS), Chennai, India, pp. 1–3. IEEE (2020)
3. Boufrioua, A.: Frequency reconfigurable antenna designs using pin diode for wireless communication applications. Wireless Pers. Commun. **110**(4), 1879–1885 (2019). https://doi.org/10.1007/s11277-019-06816-x
4. George, R., Kumar, S., Gangal, S.A., Joshi, M.: Frequency reconfigurable pixel antenna with PIN diodes. Progr. Electromagn. Res. Lett. **86**, 59–65 (2019)
5. Chen, Y., et al.: Frequency reconfigurable circular patch antenna with an arc-shaped slot ground controlled by PIN diodes. Int. J. Antennas Propag. 1–7 (2017)
6. Sun, M., Zhang, Z., Zhang, F., Chen, A.: L/S Multiband frequency-reconfigurable antenna for satellite applications. IEEE Antennas Wirel. Propag. Lett. **18**(12), 2617–2621 (2019)
7. Qin, J., Fu, X., Sun, M., Ren, Q., Chen, A.: Frequency reconfigurable antenna based on substrate integrated waveguide for S-band and C-band applications. IEEE Access **9**, 2839–2845 (2021)
8. Yuan, Y., Sun, X., Zhang, Y., Zhao, L., Wang, Y.: Design of frequency reconfigurable antenna based on metasurface. In: 2021 7th International Conference on Computer and Communications (ICCC), pp. 2165–2169. IEEE (2021)
9. Hashim, Z., Gupta, N., Prakash, A., Tripathi, R.: Dual-band UWB bandpass filter with triangular DB-DGS for WLAN applications in DSRC band. AEU-Int. J. Electron. Commun. **86**, 77–85 (2018)
10. Gottapu, S.K., Vallabhaneni, P.: Wireless sensor network localization in 3D using steerable anchors' antennas. In: Conference on Signal Processing and Communication Engineering Systems (SPACES). IEEE (2018)
11. Ramahatla, K., Mosalaosi, M., Yahya, A., Basutli, B.: Multiband reconfigurable antennas for 5G wireless and CubeSat applications: a review. IEEE Access **10**, 40910–40931 (2022)
12. Ghaffar, A., Li, X.J., Seet, B.-C., Awan, W.A., Hussain, N.: Compact multiband frequency reconfigurable antenna for 5G communications. In: 29th International Telecommunication Networks and Applications Conference (ITNAC), pp. 1–3. IEEE (2019)
13. Ullah, S., et al.: A compact frequency and radiation reconfigurable antenna for 5G and multi-standard sub-6 GHz wireless applications. Wirel. Commun. Mob. Comput. **2022**, 12 (2022). Article ID 4658082
14. Rahayu, Y., Reyhan, M.S.F., Pradana, Y.B., Kurniawan, A.: Frequency reconfigurable 5G rectangular patch antenna. In: 26th IEEE Asia-Pacific Conference on Communications (APCC), pp. 19–22. IEEE (2021)
15. Chen, Y., Liu, Q., Zhang, Y., Li, H., Zong, W.: Design of a compact base station antenna for 5G N78-band application. In: IEEE 3rd International Conference on Electronic Information and Communication Technology (ICEICT), pp. 229–231. IEEE (2020)
16. Hassan, W.M., Ibrahim, K.M., Attiya, A.M.: MIMO antenna for N48, N77, N78 5G applications. Progr. Electromagn. Res. C **117**, 129–143 (2021)

Analysis of Acoustic Channel Model Characteristics in Deep-Sea Water

Ch. Venkateswara Rao[1(✉)], M. Ravi Sankar[2], P. S R Charan[1], V. Bavya Sri[1], A. Bhaavan Sri Sailesh[1], M. Sri Uma Suseela[1], and N. Padmavathy[1]

[1] Department of ECE, Vishnu Institute of Technology, Bhimavaram 534202, India
venkateswararao.c@vishnu.edu.in

[2] Department of ECE, Sasi Institute of Technology and Engineering, Tadepalligudem, India

Abstract. Undersea acoustic communications have drawn a lot of attention recently as their uses start to transition from military to commercial. The acoustic properties of the ocean are characterized by their tremendous complexity and dynamic nature. The parameters such as; depth, temperature, salinity, location, time of day, and season of the underwater medium influences the acoustic signal propagation. However, these medium parameters are varying arbitrarily depending upon shallow and deep-water divisions of the ocean. In addition to the medium parameters, the characteristics of the acoustic channel (transmission loss, absorption and multi-path) are affected by variation in the acoustic signal speed in underwater. The influence of the aforementioned parameters alters the velocity of acoustic transmission, which affects network connectivity. Because research in the undersea environment is expanding rapidly, proficient channel modelling is required to demonstrate the effect of sound speed variations with respect to medium parameters. As a result, an acoustic channel has been modelled in this work, which analyses the sound speed variation in deep water with respect to underwater medium parameters. In addition, the proposed model evaluates absorption and transmission losses in a deep-water scenario.

Keywords: Absorption · Acoustic Channel · Deepsea · Sound Speed · Temperature · Transmission Loss · Salinity

1 Introduction

In the scientific community, underwater acoustic sensor networks (UASN) are becoming more significant due to their function as enabling technology for a variety of applications. The underwater acoustic sensor networks (UASN) are composed of submerged sensors and are used to gather information on the parts of rivers and oceans that have not previously been explored. These networks are dispersed across a study region and contain a configurable number of vehicles, anchored sensors, and floating sensors [1]. These nodes establish a communication channel with one or more hops between them. Optic, radio, electromagnetic, and acoustic waves are all capable of being used by underwater equipment to communicate with one another [2]. Since it can transmit digital data

N. Gupta et al. (Eds.): IC4S 2022, LNICST 472, pp. 234–243, 2023.
https://doi.org/10.1007/978-3-031-28975-0_20

through an underwater channel and has a large range, acoustic communication is the preferred method among researchers. The network characteristics of the UASN (node mobility, transmission range, power), as well as the characteristics of the underwater medium (temperature, pressure, salinity, and pH), introduce significant research challenges, such as limited bandwidth, multipath fading, limited battery, and limited data capacity [3].

The main factors that matter in an underwater environment are temperature, salinity, and sound speed with respect to depth [4]. Due to a variety of characteristics, such as wave height, turbid currents, water pressure, water chemical compositions, and wave speed, the undersea environment is also unpredictable [5]. The proposed channel model must be able to recognize changes in underwater medium characteristics and network parameters in order to build a reliable network. Usually, in underwater, the acoustic channel model performance is influenced by its operating frequency and the speed sound [6]. Temperature, salinity, depth, and pH are all factors that affect sound speed in underwater [7]. The sound speed changes as temperature and salinity change with regard to depth in the water. Seawater salinity varies according to both water depth and geographic location. Salinity have been measured as a part per thousand and calculated from the ocean dissolved salt concentrations (ppt) [8–10].

In Polar Regions, salinity is less than 30 ppt, despite the fact that the average salinity is between 31 and 37 ppt. On the other hand, salinity, temperature, transmission distance, and frequency of operation all affect how well sound travels through sea water [11]. These irregular differences in sound speed (depending on temperature, salinity, water depth, and pH) and absorption affect the formation of linkages between the sensor nodes (dependent on acoustic frequency, transmission distance). The connectedness of UASN also depends on the development of efficient communication links between the sensor nodes. An effective acoustic channel model that takes into account the effects of sound speed, absorption losses, and transmission losses must be included in order to provide reliable communication between the sensor nodes in a deep underwater network.

2 Related Work

The marine environment has a significant impact on wireless acoustic signal propagation across a water body as the channel of propagation. The Doppler shift, strong multi-path propagation, high attenuation, constrained bandwidth, severe fading, lengthy delay spread, fast time variation of the channel, route loss, and noise are only a few of the challenges the acoustic channel faces [12]. Therefore, in order to build and create efficient underwater communication systems, it is essential to do research and have a thorough understanding of how the underwater environment affects the communication signal in both the regions (shallow and deep water) of marine. Even though, the terrestrial networks have attained significant attention in the research groups, the implementations of these networks in underwater is a challenging task due to excessive absorption of electromagnetic waves in underwater. In addition, the terrestrial networks are differed from underwater in many network constraints [13–24] such as; propagation delay, topology, power needs, mobility, network life, data rate, etc. The foremost parameter that affects the performance of acoustics is variable sound speed. Usually, the temperature changes have been reduced in Deepsea water locations when compared to shallow water. The authors in [12] have provided and investigated the underlying physics of fundamental wave propagations using the fundamental physics concepts before contrasting the problems and effects of using various communication carriers (acoustic, EM, and optical).

The authors in [25] has explored the effects of propagation characteristics on underwater communication, including sound speed, channel latency, absorption, scattering, multipath, waveguide effects, and ambient noise. The authors in [26] used increased propagation loss and ambient noise models to examine the dependence of the channel capacity on depth and temperature. For reducing the mistakes in the sound speed in oceans and seas, a mathematical model [27] that offers the conversion of atmospheric pressure to depth and depth to atmospheric pressure has been presented. In, [28] authors have described an experimental setup that shows how salinity, temperature, and pressure affect the physical properties of the deployed environment and sound speed fluctuations. A method [29] for increasing the localization accuracy in the water by estimating the sound speed at a specific spot with time. For network simulation in underwater environments, an acoustic channel model has been developed [30]. A real-time measurement of the route loss for an underwater acoustic channel has been made [31]. To improve accuracy and throughput, modelling underwater acoustic channel characteristics has been applied recent breakthroughs in deep learning and artificial intelligence. To increase the accuracy of channel models, a deep learning-based framework for underwater channel modelling has been presented [32]. The main statistical characteristics of the channel model have been outlined and examined in [33].

The literature has shed light on how changes in the properties of the underwater medium affect underwater sound propagation. The impact of temperature, salinity, absorption, and transmission losses caused by changes in the sound speed must therefore be addressed and taken into account. The link development in the network is altered by these fluctuations in sound speed caused by the unpredictability of the undersea environment. As a result, the proposed work objective is to analyses the acoustic channel properties in the Deepsea water region while taking transmission losses and absorption into account.

3 Channel Model

An acoustic channel that considers the effects of temperature, salinity, absorption, and transmission losses on sound speed has been designed for simulating an underwater acoustic channel in shallow and Deepsea water. In a Deepsea water scenario, the sound speed has first been measured at various depths by adjusting salinity and temperature, and then the losses resulting from the change in sound speed have been determined. In addition, the transmission losses and absorption with respect to frequency and distance has been evaluated for deep water scenario.

3.1 Sound Velocity

The incredibly slow pace at which sound moves through water is one of the main differences between acoustic waves and EM transmission. Undersea sound speed is influenced by conditions including pressure, salinity, and temperature. The speed of sound near the ocean surface is normally around 1520 m/s, which is four orders of magnitude faster than the speed of sound in air but five orders of magnitude slower than the speed of light. The sound speed in water is directly impacted by environmental changes (temperature, salinity, and depth). Mackenzie provides an empirical formula [34] for calculating sound velocity as a function of temperature, salinity, and depth is represented using (1).

$$c(T,S,z) = a_1 + a_2T + a_3T^2 + a_4T^3 + a_5(S-35) + a_6z + a_7z^2 + a_8T(S-35)a_9Tz^3 \quad (1)$$

3.2 Propagation Loss of Sound

Spreading, dispersion, and absorption all contribute to the attenuation in undersea acoustic signals. Spreading loss is a metric for signal deterioration brought on by the geometrical spreading effect that occurs as a sound wave moves away from its source. Cylindrical and spherical spreading in underwater acoustics, are two different types of spreading mechanisms. For cylindrical and spherical spreading the loss is expressed using (2) and (3) respectively. L_{CS} represents loss due to cylindrical spreading, L_{SS} represents loss due to spherical spreading and R_t represents the transmission range [35].

$$L_{CS} = 10\log(R_t) \quad (2)$$

$$L_{SS} = 20\log(R_t) \quad (3)$$

3.3 Absorption Coefficient

For the frequency range of 100 Hz to 1 MHz, an empirical formula [35] for the absorption coefficient has been expressed using (4). It changes with frequency, pressure (depth), and temperature.

$$\alpha = \frac{A_1P_1f_1f^2}{f^2+f_1^2} + \frac{A_2P_2f_2f^2}{f^2+f_2^2} + A_3P_3f^2 \quad (4)$$

3.4 Absorption Loss

The absorption loss, which is range-dependent and calculated using (5), represents the energy loss of sound as a result of the transformation of energy into heat due to the chemical features of viscous friction and ionic relaxation in the ocean.

$$L_{ab} = (\alpha \times R_t) \times 10^{-3} \quad (5)$$

3.5 Transmission Loss

Transmission loss [36], which is defined as the cumulative loss of acoustic strength caused by an acoustic pressure wave as it moves away from its source which is represented by using (6).

$$TL_{Deepsea} = L_{SS} + L_{ab} \tag{6}$$

4 Implementation Parameters

The parameters considered for analyzing the transmission loss, absorption loss, spreading loss and sound speed has been listed in Table 1.

Table 1. Implementation Parameters

Parameter	Range
Depth	100–8000 m
Temperature	30–4 °C
Salinity	30–37 ppt
Frequency	100 Hz–100 kHz
pH	7.8
R_t	100 m

5 Simulation Results

The acoustic medium is the primary means of wireless data communication in marine environment, and the speed of sound is the most fundamental characteristic that influences the data rates that can be achieved, as well as the quality of service, latency, and other crucial network parameters in this channel. The sound speed has unpredictable variations in underwater due to abnormal changes in temperature, salinity, depth and pH with respect to the season, time and location of the ocean.

In addition, the sound speed also influenced by numerous factors, including wave height, turbid currents, water pressure, chemical compositions of the water, and wave speed. Whereas, the variation in aforementioned parameters has different profiles in shallow and deep-water divisions of the ocean. The sound speed profile exhibits abnormal variations in shallow water because of drastic change in temperature gradients of the water column across the water depth. Whereas, the temperature is almost constant (4 °C) at the deep water scenarios, where the sound speed profile has minimal variations. It is clearly depicted in Fig. 1, that the sound speed is varying with temperature and depth. When the depth is extended to 7000 m and the temperature is lowered to 4 °C, the sound

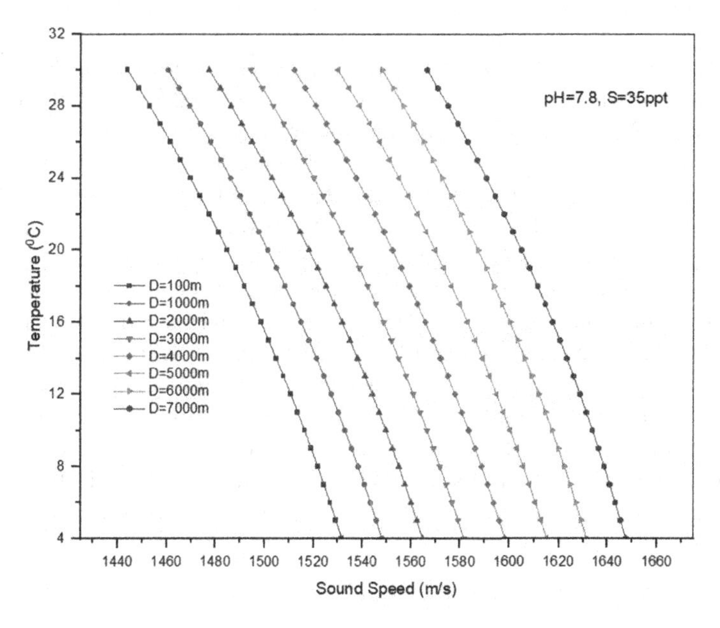

Fig. 1. Effect of Temperature variations on sound speed in deep-water

speed increases to 1650 m/s (see Fig. 1) from the initial value of 1450 m/s at a certain temperature and depth ($T = 30$ °C, $D = 100$ m). Similarly, salinity of the ocean water

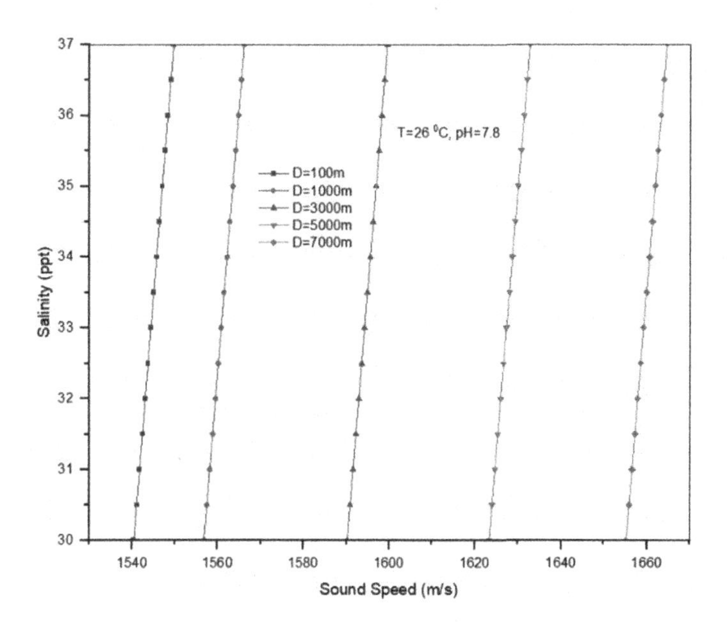

Fig. 2. Effect of salinity variations on sound speed in deep-water

increases along the depth, which also influences the sound speed in deep water. This is clearly depicted in Fig. 2, that the sound speed increases with increase in depth as well as salinity. At a particular salinity (S = 33 ppt), the sound speed attained different profiles (varying from 1540 m/s to 1650 m/s) along the depth (see Fig. 2).

Absorption, which results from the transformation of acoustic energy into heat, is the principal cause of attenuation. As the distance and frequency rise, the attenuation grows. The influence of different absorption due to various chemical compositions of the underwater medium has been depicted in Fig. 3. It is clear that boric acid (H_3BO_3), magnesium sulphate ($MgSO_4$), and pure water are the main contributors to attenuation at frequencies below 1 kHz, between 1 kHz and 100 kHz, and above 100 kHz, respectively (see Fig. 3). It is also obvious that the orders of magnitude vary widely and that attenuation rises sharply with frequency. For frequencies of 1 kHz and below, attenuation is less than a few hundredths of a dB/km; hence, it is not a limiting factor. Approximately 1 dB/km of attenuation occurs at 10 kHz, restricting ranges of more than a few tens of kilometres. The transmission losses are frequency and range dependent. According to Fig. 4, transmission losses increase as frequency increases, while transmission losses decrease as depth increases.

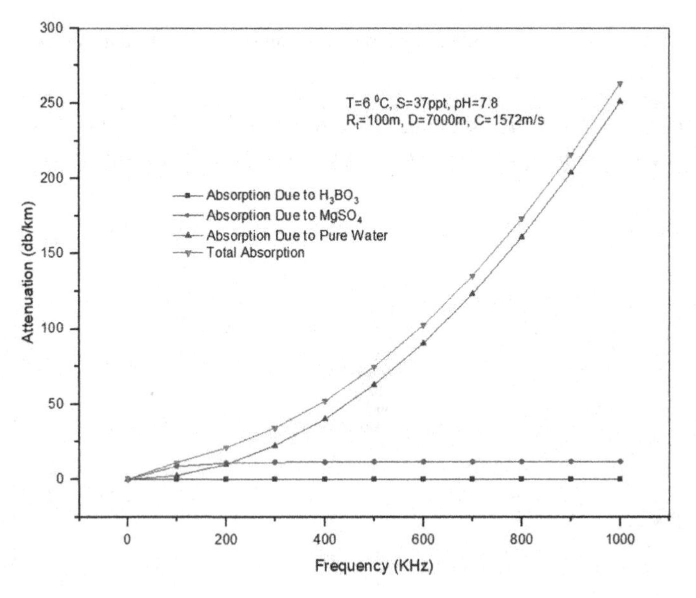

Fig. 3. Attenuation in deep water due to absorption

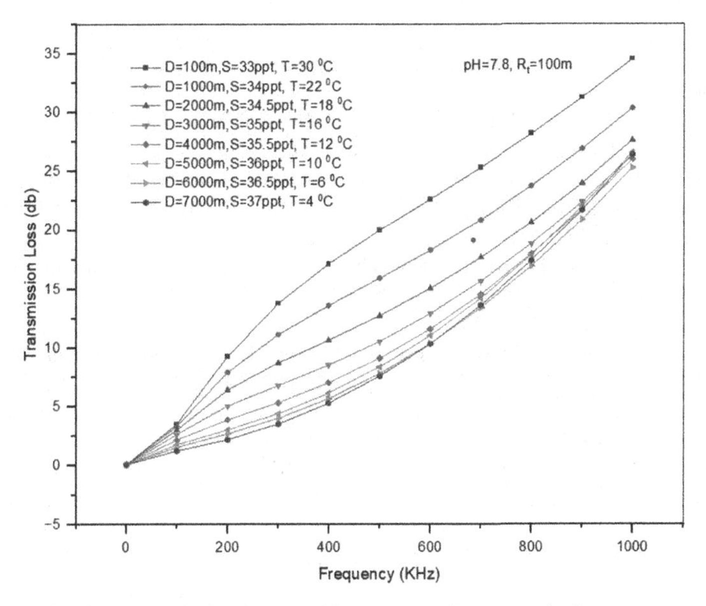

Fig. 4. Transmission losses with respect to frequency in deep-water

6 Conclusion

This study proposes an acoustic channel model that investigates the effect of underwater medium factors such as temperature and salinity on sound speed. By varying different depths in a deep-water scenario, the proposed channel model investigates the effect of salinity and temperature at a fixed pH. The proposed channel model also investigates the effect of absorption due to different chemical compositions of water with respect to the frequency of the acoustic signal. Finally, the transmission losses have been estimated at various depths in deep-water with respect to the frequency at a fixed transmission range. The simulation results show that transmission and absorption losses are frequency dependent. These two losses have increased in proportion to the frequency. As the temperature and salinity decrease gradually and along the depth, the sound speed increases in deep-water.

References

1. Sozer, E.M., Stojanovic, M., Proakis, J.G.: Underwater acoustic networks. IEEE J. Ocean Eng. **25**(1), 72–83 (2000)
2. Akyildiz, I.F., Pompili, D., Melodia, T.: Underwater acoustic sensor networks: research challenges. Ad Hoc Netw. **3**(3), 257–279 (2005)
3. Barbeau, M., Garcia-Alfaro, J., Kranakis, E., Porretta, S.: The sound of communication in underwater acoustic sensor networks. In: Zhou, Y., Kunz, T. (eds.) Ad Hoc Networks. LNICSSITE, vol. 223, pp. 13–23. Springer, Cham (2018). https://doi.org/10.1007/978-3-319-74439-1_2

4. Akyildiz, I.F., Pompili, D., Melodia, T.: Challenges for efficient communication in underwater acoustic sensor networks. ACM SIGBED Rev. – Spec. Issue Embed. Sens. Netw. Wirel. Comput. **1**(2), 3–8 (2004)
5. Stojanovic, M., Preisig, J.: Underwater acoustic communication channels: propagation models and statistical characterization. IEEE Commun. Mag. **47**(1), 84–89 (2009)
6. Jindal, H., Saxena, S., Singh, S.: Challenges and issues in underwater acoustics sensor networks: a review. In: International Conference on Parallel, Distributed and Grid Computing Solan, pp. 251–255 (2014)
7. Ismail, N.-S., Hussein, L., Syed, A., Hafizah, S.: Analyzing the performance of acoustic channel in underwater wireless sensor network. In: Asia International Conference on Modelling & Simulation, pp. 550–555 (2010)
8. Wanga, X., Khazaiec, S., Chena, X.: Linear approximation of underwater sound speed profile: precision analysis in direct and inverse problems. Appl. Acoust. **140**, 63–73 (2018)
9. Ali, M.M., Sarika, J., Ramachandran, R.: Effect of temperature and salinity on sound speed in the central Arabian sea. Open Ocean Eng. J. **4**, 71–76 (2011)
10. Kumar, S., Prince, S., Aravind, J.V., Kumar, G.S.: Analysis on the effect of salinity in underwater wireless optical communication. Mar. Georesour. Geotechnol. **38**(3), 291–301 (2020)
11. Hovem, J.: Underwater acoustics: propagation, devices and systems. J. Electro Ceram. **19**, 339–347 (2007)
12. Chitre, M., Shahabodeen, S., Stojanovic, M.: Underwater acoustic communications and networking: recent advances and future challenges. Mar. Technol. Soc. J. **42**, 103–116 (2008)
13. Chaturvedi, S.K., Padmavathy, N.: The influence of scenario metrics on network reliability of mobile ad hoc network. Int. J. Performability Eng. **9**(1), 61–74 (2013)
14. Venkata Sai Kumar, B., Padmavathy, N.: A hybrid link reliability model for estimating path reliability of mobile ad hoc network. Procedia Comput. Sci. **171**, 2177–2185 (2020). https://doi.org/10.1016/j.procs.2020.04.235
15. Cook, J.L., Ramirez-Marquez, J.E.: Two-terminal reliability analyses for a mobile ad hoc wireless network. Reliab. Eng. Syst. Saf. **92**(6), 821–829 (2007)
16. Padmavathy, N., Chaturvedi, S.K.: A systematic approach for evaluating the reliability metrics of MANET in shadow fading environment using Monte Carlo simulation. Int. J. Performability Eng. **12**, 265–282 (2016)
17. Padmavathy, N., Chaturvedi, S.K.: Reliability evaluation of capacitated mobile ad hoc network using log-normal shadowing propagation model. Int. J. Reliab. Saf. **9**(1), 70–89 (2015)
18. Venkata Sai, B., Padmavathy, N.: A systematic approach for analyzing hop count and path reliability of mobile ad hoc networks. In: International Conference on Advances in Computing, Communications and Informatics, pp. 155–160 (2017)
19. Padmavathy, N., Anusha, K.: Dynamic reliability evaluation framework for mobile ad-hoc network with non-stationary node distribution. In: Communication and Computing Systems. CRC Press, Taylor and Francis (2018)
20. Padmavathy, N., Teja, J.R.C., Chaturvedi, S.K.: Performance evaluation of mobile ad hoc network using Monte Carlo simulation with failed nodes. In: 2nd International Conference on Electrical, Computer and Communication Technologies, pp. 1–6 (2017)
21. Cook, J.L., Ramirez-Marquez, J.E.: Reliability analysis of cluster-based ad-hoc networks. Reliab. Eng. Syst. Saf. **93**(10), 1512–1522 (2008)
22. Padmavathy, N.: An efficient distance model for the estimation of the mobile ad hoc network reliability. In: Gunjan, V.K., Garcia Diaz, V., Cardona, M., Solanki, V.K., Sunitha, K.V.N. (eds.) ICICCT 2019, pp. 65–74. Springer, Singapore (2020). https://doi.org/10.1007/978-981-13-8461-5_8

23. Venkateswara Rao, C., Padmavathy, N.: Effect of link reliability and interference on two-terminal reliability of mobile ad hoc network. In: Verma, P., Charan, C., Fernando, X., Ganesan, S. (eds.) Advances in Data Computing, Communication and Security. LNDECT, vol. 106, pp. 555–565. Springer, Singapore (2022). https://doi.org/10.1007/978-981-16-8403-6_51
24. Rao, C.V., Padmavathy, N., Chaturvedi, S.K.: Reliability evaluation of mobile ad hoc networks: with and without interference. In: IEEE 7th International Advance Computing Conference, pp. 233–238 (2017)
25. Preisig, J.: Acoustic propagation considerations for underwater acoustic communications network development. Mob. Comput. Commun. Rev. **11**(4), 2–10 (2006)
26. Sehgal, A., Tumar, I., Schonwalder, J.: Variability of available capacity due to the effects of depth and temperature in the underwater acoustic communication channel. In: OCEANS 2009-EUROPE, Bremen, pp. 1–6 (2009)
27. Leroy, C.C., Parthiot, F.: Depth-pressure relationships in the oceans and seas. J. Acoustic Soc. Am. **103**(3), 1346–1352 (1998)
28. Yuwono, N.P., Arifianto, D., Widjiati, E., Wirawan: Underwater sound propagation characteristics at mini underwater test tank with varied salinity and temperature. In: 6th International Conference on Information Technology and Electrical Engineering (ICITEE), pp. 1–5 (2014)
29. Shi, H., Kruger, D., Nickerson, J.V.: Incorporating environmental information into underwater acoustic sensor coverage estimation in estuaries. In: IEEE Military Communications Conference, MILCOM 2007, pp. 1–7 (2007)
30. Morozs, N., Gorma, W., Henson, B.T., Shen, L., Mitchell, P.D., Zakharov, Y.V.: Channel modeling for underwater acoustic network simulation. IEEE Access **8**, 136151–136175 (2020)
31. Lee, H.K., Lee, B.M.: An underwater acoustic channel modeling for internet of things networks. Wirel. Pers. Commun. **116**(3), 2697–2722 (2020). https://doi.org/10.1007/s11277-020-07817-x
32. Onasami, O., Adesina, D., Qian, L.: Underwater acoustic communication channel modeling using deep learning. In: 15th International Conference on Underwater Networks & Systems (WUWNet 2021), China (2021)
33. Zhu, X., Wang, C.-X., Ma, R.: A 2D non-stationary channel model for underwater acoustic communication systems. In: IEEE 93rd Vehicular Technology Conference (VTC2021-Spring), pp. 1–6 (2021)
34. Mackenzie, K.V.: Nine-term equation for sound speed in the oceans. J. Acoust. Soc. Am. **70**(3), 807–812 (1981)
35. Etter, P.C.: Underwater Acoustic Modeling and Simulation, 3rd edn. Spon Press, New York (2003)
36. Padmavathy, N., Venkateswara Rao, Ch.: Reliability evaluation of underwater sensor network in shallow water based on propagation model. In: Journal of Physics: Conference Series, vol. 1921, p. 012018, 1–17 (2021)

A Comprehensive Review on Channel Estimation Methods for Millimeter Wave MIMO Systems

Ch. V. V. S. Srinivas[1,2](✉) and Somasekhar Borugadda[3]

[1] ECE Department, Andhra University, Visakhapatnam, India
srinuch17@gmail.com
[2] ECE Department, Vishnu Institute of Technology, Bhimavaram, India
[3] Anil Neerukonda Institute of Technology and Sciences, Visakhapatnam, India

Abstract. The major requirements of fifth-generation (5G) cellular networks and further are low latency and great speed of data transmission. One such way to fulfill these requirements is to make use of 5G enabling technology called Millimeter-wave (mmWave) communication. The cellular systems relying on this technology can function with higher data transfer rate up to gbps due to the availability of large band width. A very high transceiver intricacy is required exorbitantly in the case of conventional MIMO methods whereas digital precoding/conjoining is generally accomplished at baseband with single radio frequency (RF) sequence and a single ADC per antenna in case of traditional MIMO systems. So an all-digital processing approach is prohibitive owing to the high price and power depletion of RF chains and ADCs at frequencies of mmWave. This appraisal primarily aims at offering a complete analysis of channel approximation techniques accompanied with various diversified millimeter wave scheme. Consequently, in terms of their corresponding aids and inadequacies a comparison is also provided among prevailing resolutions.

Keywords: Millimeter-wave communications · 5G cellular systems · Massive MIMO · Hybrid architecture · Channel estimation · Beamforming · Compressive sensing

1 Introduction and Background

Technology has emerged all its way to the next level with the upbringing of various evolving applications like AR, VR, MR, holography projection, high definition videos, smart cities, industrial IoT, connected cars and many more. The high ended technologies mention above need to link devices immensely and they need to transmit the data with high speed and also more data need to be exchanged between devices. So we go for a fifth generation technology in which the user data rate is increased by a factor of 10–100 (up to 10 Gigabps), latency is reduced by factor 10, connectivity density is increased by 10 when compared to 4G, and also to decrease the price and power intake [1]. Owing to this immensely growing ultimatum for data congestion and enormous connectivity and

N. Gupta et al. (Eds.): IC4S 2022, LNICST 472, pp. 244–255, 2023.
https://doi.org/10.1007/978-3-031-28975-0_21

in addition the inadequacy in the sub-6 GHz radio spectrum, investigators are attempting to put forward innovative elucidations. These are primarily centered either on making the network busy or on developing extra bands of frequency or on employing the novel schemes of signal processing.

In terms of employing several bands, the millimeter wave frequency in between the range of 30 Giga Hertz and 300 Giga Hertz (i.e. range of 1–10 mm wave lengths), and the vast underutilized bandwidth in these bands will let the wireless systems to handle massive increases in capacity ultimatum as the employed bandwidth increases. Because of this reason, mmWave communications play a superior role in fifth and further peers of cellular networks [2].

The basic challenge with the mm Wave communication systems is that they have a difficulty of large path loss of free space when related to the ones of sub-6 Giga Hertz although their bandwidth is striking. Adding to that, in some mm Wave bands, a significant attenuation is witnessed due to atmospheric changes, and extreme weather conditions like the effects of rain and snow. Consequently, in such cases, the mm Wave communications may be appropriate only for indoor applications i.e., very close distance communications since it is unsafe to broadcast over a small distance of meters. For outdoor mobile communications, by either raising the power of transmission or by employing high-gain, high-directional antennas, one can overcome this difficulty and owing to this, a better range of transmission is expected.

The high transmission power is possible by facilitating constricted steering beams in mm Wave systems, since the transmission power is always restricted by conventions. A lot much gain is essential in the desirable direction and lesser gain in the undesirable region. These can be accomplished by allowing the transmitter and the receiver to direct towards each other. Signal processing technique such as beam forming is employed to attain the requisite high directivity, which is possible by taking the number of antennas in each antenna group at the sender and receiver ends huge. This method is enabled by the lesser mm Wave signal wave length of which makes it feasible to hold together more antennas while overcome maintaining the size of the array small.

One more feature of millimeter wave structure design emanates from the unfeasibility to straight away smear the outmoded digital transceiver designs, which is engaged in sub-6 Giga Hertz, straight to the millimeter wave structures since there is large power intake of mm Wave RF chains. In order to solve this problem, of late, some novel personalized multiple-input-multiple-output (MIMO) structural designs i.e. *Totally-analog fusion* and *few bit Analog to Digital Converters (ADC)* designs were recommended. Whichever architecture is implemented, the goal is to reduce the total amount of power consumed by minimizes the Radio Frequency chains count or the power consumed per unit.

Channel estimation is the most essential part of any telecommunication structure as it is very much required for augmenting the performance of the link. But because of the multifaceted architecture of transceiver, massive number of antennas and large misused bandwidth, attaining Channel State Information (CSI) is quite a bit challenge predominant in millimeter wave systems. The main intention of this appraisal is to put forward a complete general idea of the prevailing channel estimation techniques for millimeter wave systems. Each method of estimation is analyzed and provided with the appropriate level of particulars in a clear and crisp manner. Also a reasonable assessment among them is carried out by a measure of structural design and performance.

In the current study, many diverse prevailing research works were considered and different assumptions were made to them. They are with regard to the: i) Type of data transmission (simplex or half duplex or full duplex), ii) Number of users of the system (whether single or many), iii) Dimensions of channel estimation (2D or 3D), iv)Type of the channel (frequency selective or not), v) mode of the channel for implementation.

Table 1. Review of millimeter wave Communications System existing Surveys.

Coverage and Access	[3]
Channel Estimation Technique	[3, 4]
Channel Measurement Technique	[3, 4, 6]
Channel Modeling Technique	[3, 4]
Cross-Layer Design	[4, 7]
Technical Potentials and Key Challenges	[2, 3]
MIMO system Architecture	[3, 5]
Performance Analysis	[5]
Propagation Characteristics of channel	[2, 5, 6]
Standardization	[4, 7]

Table 1 illustrates an examination of some attributes associated with mmWave systems and the appraisals in which respective area was fundamentally gone through.

2 Characteristics of mmWave Communications

For applications like radar and dedicated communication, mmWave communications have been utilized previously. Later, IEEE has developed certain standards such as IEEE 802.15.3c [9] for wireless PAN's, IEEE 802.11ad [10] for wireless LAN's and Wireless HD [11] for wireless HDMI in the last decade. Of late, the mobile link exploration communal has paid much attention to sub-100 GHz structures functioning in the bands of 28, 38, 71 and 81 GHz, where as the band over 100 GHz has only been explored by a limited latest article [12]. At mmWave frequencies, the propagation characteristics distinct from those of a standard sub 6 GHz system.

A. *Propagation Characteristics*

The free space path loss is a significant characteristic of propagation where a clear comparison is made between sub-6 GHz and mm Wave systems and it is given by the following expression.

$$PL_{FS} \propto \frac{d^n}{\lambda^2} \quad (1)$$

where *FS* stands for free space and n is the path loss coefficient, which is usually equivalent to 2. But, in certain circumstances of the cellular networks and indoor routines n is smaller than 2. In some cases such as in the course of extreme weather conditions while propagation, the value of the path loss coefficient could come near the value of 6 [14], λ is the wavelength and d is the separation distance between the transmitter and receiver [15]. An analysis is made in terms of distance of transmission, antenna array, propagation environment between microwave broadcast in the 1.8 Giga Hertz Global System for Mobile (GSM) Communication band and mmWave propagation in the 73 Giga Hertz bands was compared, and found an additional loss of 32 dB.

Path loss includes the reduction in strength of mm Wave signals which are caused due to atmospheric changes principally by O_2 and water vapor absorption, along with the scattering effects of the rain. The atmospheric attenuation is proportional to the operating frequency. At 60 and 120 GHz, atmospheric oxygen absorption is very severe, while water vapour absorption is generally very strong at 180 GHz [16].

B. *Technical Potential*

Owing to the high bandwidth efficiency and shorter wavelengths, the 5G cellular technology gives the mm Wave Communication systems the front face. Of course there are a few characteristics of mmWave channel which were quite a bit challenging, but they can be over written because of the presence of some benefits.

Large Bandwidth: Concentration restricted alone on the profoundly engaged traditional sub-6 Giga Hertz frequency bands doesn't serve the purpose of achieving very high speeds of data. This high data rates can be attained by providing a very large bandwidth. So the large number of mmWave bands is made available to achieve enormous data exchange in spite of low spectral proficiency [17].

Short Wavelength: Antennas with high directivity can be employed for overcoming the problem of shorter wavelengths finally resulting in the free space path loss. This can be overcome by spending on antenna arrays and methods of beamforming. Narrower beams involve higher antenna count in the array which is attuned with mmWave structures.

III. *Technologies that enable*

Millimeter Wave communications tools is enabled by other techniques like massive MIMO, signal processing and advancements in circuit plan and amalgamation.
Massive MIMO: It is a most essential method for improving the capacity of cellular networks. Massive MIMO employs a large number of antennas at base stations and mobile terminals. So far the traditional sub-6 GHz systems have made use of this Massive MIMO, and so it is as well very much important for mmWave structures where more amount of directivity is obligatory [18]. Due to the availability of higher frequency bands, antenna arrays were designed with a more number of antennas.

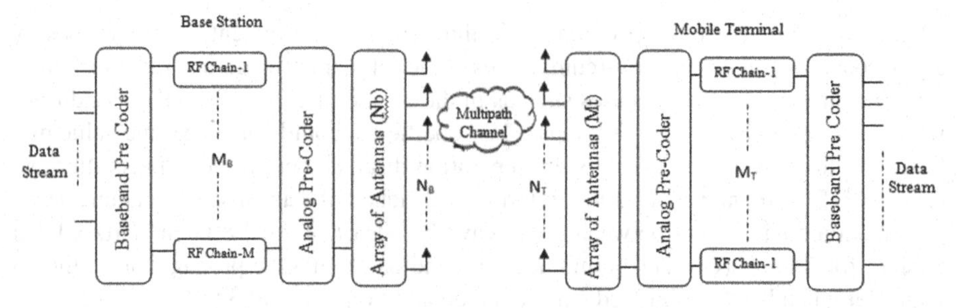

Fig. 1. Overall mmWave transceiver system architecture.

Enhanced Signal Processing Techniques: The problem with the traditional MIMO systems is that they are made totally digital and a single RF chain is made available to each antenna. Due to this the difficulties such as high enactment cost and high energy in take arise [19, 20]. Certainly, it has been demonstrated that Radio Frequency constituents can ingest up to 70 percent of the whole transceiver power intake [21]. So instead of going after the completely digitalized one, quite a few mmWave hybrid designs were put forward.

3 Massive MIMO for Millimeter Wave Systems

For millimeter wave communication systems, more number of antennas must be used, both at the transmitter and receiving ends. As per the previous exploration carried out, it is recommended that the base station have to be possessed with 32 to 256 antennas and the mobile terminal with 4 to 16 antennas [22].

The overall design of hybrid mm Wave transceiver is depicted by means of a block diagram displayed in Fig. 1. Without loss of generality, a Base Station which is constructed with number of Radio Frequency chains (M_B) and a number of antennas (N_B) then ($M_B < N_B$) is expected to fortified with M_T Radio Frequency chains and N_T antennas ($M_T < N_T$) so as to interchange some blocks of data [5]. In general, the count of Radio Frequency chains at base stations is more than that of mobile terminals. The same design might be drawn-out to several numbers of handlers. However, here, a point-to-point transmission is considered.

The hybrid design's baseband digital processing and analogue circuit work together to generate beams for the sender or/and receiver. In disparity by the outmode overall-digital design that would not contemplate an analog circuit beforehand the Radio Frequency chains at the sender and contemplates a single Radio Frequency chain per every antenna, i.e. $M_B = N_B$ and $M_T = N_T$ [3]. Because of the problems mentioned above, nowadays a complete digital design is sidestepped in millimeter wave communication system design. In the place of that novel hybrid designs were proposed so as to make the massive MIMO system achievable. The awareness is to lessen the total power intake and the total amount needed out which can be made possible by diminishing the count of Radio Frequency chains and the resolution of the Analog to Digital (ADC) Converters.

Hybrid Style: The hybrid design characterizes a negotiation between an entirely digital structural design (with the same number of antennas and Radio Frequency chains) and a completely analogue design (where only single Radio Frequency chain is utilized). Both analogue and digital domains are used for precoding (at transmitter side) and combining (at the receiver side). The aim is to make utilization of a dimension compact pre-coder/combiner with a very less count of Radio Frequency chains while still banking on complete-size analog pre-coder/combiner. This hybrid architecture has enhanced the performance in terms of reduction in the RF chain count but the results are shown to be not very much enhanced from complete digital design [8].

mmWave MIMO Channel Modeling: Physical modeling and analytical modeling are the two different types of mmWave channel architectures. The physical models are centered on the electromagnetic properties of signal broadcast between arrays of antennas matching to transmitter and receiver. They are an excellent choice for mmWave MIMO channels because they can effectively redirect to the measured parameters and are well-known. The latter one i.e., the analytical models on the other hand, are more suitable for the development of algorithm and for scrutinizing the system.

Table 2. Techniques for estimating channel in hybrid architecture systems

Reference paper	Method	No. of Users	Up Link/Down Link	Description
Alkhateeb et al. [23]	Divide and conquer approach	Single	Down Link	Orthogonal Matching Pursuit, Least Square Estimation
He et al. [25]	Mode by mode approach	Single	Up Link/Down Link	Non Line of Sight channel based on the Time Division Duplexing correlation statistics
Lee et al. [24]	Open loop system	Single	Not specified	Orthogonal Matching Pursuit, MG-OMP, LSE
Schniter et al. [26]	Aperture shaping	Not specified	Not specified	Least Absolute Shrinkage and Selection Operator, LMMSE

(continued)

Table 2. (*continued*)

Reference paper	Method	No. of Users	Up Link/Down Link	Description
Payami et al. [27]	Ping pong approach	Single	Not specified	The amount of multi-path components has no effect on the training time
Peng et al. [29]	Antenna Array with Virtual Elements	Single	UpLink	The angular estimate resolution is improved using a CS-based approach
Kokshoorn et al. [28]	Overlapped beam patterns	Single	Not specified	MRC, used to track fast changing channels
Montagner et al. [30]	2D Discrete Fourier Transform	Single	Not specified	DFT, iterative cancellation method
Mendezrial et al. [31]	Switches	Single	Down Link	Orthogonal Matching Pursuit, M-OMP
Han et al. [32]	Two stage asymmetric	Multi	Down Link	Exhaustive search, Compressive Sensing
Park et al. [33]	Spatial covariance Technique	Single	Up Link	Orthogonal Matching Pursuit, S-OMP, C-OMP, Dynamic Simultaneous -OMP, Dynamic Covariance OMP
Guo et al. [34]	Dimension Deficient	Single	UpLink	CoSaMP, it reduce the accidental errors

The channel model communicates the NTX NB composite matrix HDL of the Narrowband Down Link (DL) channel is expressed as,

$$H_{DL} = \sqrt{\frac{N_B N_T}{L\alpha_{PL,DL}}} \sum_{l=1}^{L} \alpha_{l,DL} a_T(\theta_{l,T}, \varphi_{l,T}) a_B^H(\theta_{l,B}, \varphi_{l,B}), \tag{2}$$

where L represents path count between the Base Station (BS) and Mobile Terminal (MT), $(\theta_{l,T}, \Phi_{l,T})$ are the azimuth angles of arrival and elevation angles of arrival (AoA) and $(\theta_{l,B}, \Phi_{l,B})$ are the azimuth angles of departure and elevation angles of departure

(AoD), $\alpha_{PL,DL}$ is the regular value of the path-loss, $\alpha_{l,DL}$ is the complex increase of gain of the lth track in the Down Link channel. The Eq. (2) ought to be revised to signify the D-delay down link channel exemplary because of the reason that the channel is frequency-discerning when the structure is extended to wide range of frequencies.

Henceforth the equation is produced as:

$$H_{DL} = \sqrt{\frac{N_B N_T}{L\alpha_{PL,DL}}} \sum_{l=1}^{L} \alpha_{l,DL} p_B(dT_s - \tau_l)\ldots$$
$$a_T(\theta_{l,T}, \varphi_{l,T}) a_B^H(\theta_{l,B}, \varphi_{l,B}), \quad (3)$$

At the Base Station, Where $p_B(.)$ is a combination of the pulse shaping filter and other filters responses.

4 Techniques for Estimating Channels in Hybrid Architecture Systems

Taking into consideration the previously mentioned restraints, a few channel approximation approaches by way of this hybrid design have been suggested. These approaches, along with others, are detailed below, and they are given in Table 2 with t corresponding references. Many of the proposals undertake a narrowband constant frequency mmWave channel prototype. Various techniques for mm wave channel estimation with hybrid architectures are presented in this section.

Divide and Conquer Approach: In this method, [23] a low-intricacy algorithm has been proposed called adaptive channel estimation for mmWave channel. For this approach, more and less number of antenna arrays and RF units are employed at both the sides of Base Station and Mobile Terminal respectively. The approximation procedure is separated into several phases. The Angle of Arrivals/Angle of Departures angular values are parted into K non-overlaid angular sub-slots at each stage, and K beam patterns are employed to direct the pilot signal and syndicate the signal at the destination. As a result, each beam arrangement at the sender is joined with K beam structures at the receiver, requiring K^2 time slots for each stage to span all possible transmit-receive beam pattern combinations. The authors constructed and offered a predetermined codebook that contains the beam patterns. Evaluating the amounts of the K^2acknowledged signals is then used to decide or surmount the subsequent Angle of Arrivals / Angle of Departures sub-range for the following stage.

Ping-pong Approaches: This algorithm customs ping-pong iterations. In order to attain the channel parameters, this algorithm entails an intricacy of $O(KL^2 N_B log_K(N_G/L))$, wherefrom an even network of NG points, the parameters AoAs and AoDs are acquired. Two-phase procedure for estimation of channel with a single-user (SU) is presented in [27] and in addition a codebook design which is analogous to that in [23] is also developed. The procedure is categorized by a dual phase handshaking amidst the sender and the receiver.

Overlapped Beam Patterns Technique: In [28] by making use of a conception of making an approximation on the beam patterns that are overlapped, a fast channel estimation algorithm is put forward. When compared to [23], this idea lessens the time turns desirable to approximate the channel by $K^2/\log_2^2(K+1)$ with a minor deprivation in performance. This minor deterioration can be tolerated when tracking rapidly changing channels is essential.

Open-loop Approach: In [24] proposed an open-loop channel approximation method that does not require a feedback loop. The procedure is grounded on CS methods and is recommended for an SU mmWave structure. Just like various channel approximation procedures, the researchers employed the Orthogonal Matching Pursuit and least squares estimation algorithm (LSE) practices to accomplish the approximation of the Angle of Arrivals/Angle of Departures increase in power correspondingly.

Mode-by-Mode Method: A different structure was proposed in [25] in order to assess a time basely correlated Non Line of Sight (NLOS) in mmWave MIMO channel. To effectively trail the channel disparities, the authors former changed the parametric channel structure into a progression time basely correlated MIMO channel model. The structure is established on TDD correlation statistics and investigates the channel's reciprocity. The mode by mode technique, a suggested procedure, updates each column of the analog pre-coder and combiner.

Aperture Shaping Technique: In [26], The aim of augmenting the sparseness of the virtual MIMO channel have recommended a procedure labeled as aperture shaping. Concisely, shaping of aperture is implemented by smearing a constant gain value at individual antennas at either ends of the Transmitter and Receiver. The shaping coefficients are augmented to get the most out of the signal-to-interference ratio. Adding to that, to further uncover the channel sparsity, They used a millimeter Wave system, which employs modulation and demodulation practices established on Fast Fourier transforms (FFT).

Antenna Array in Virtual Element Approach: The authors in [29] suggested a novel idea, which outspreads the actual antenna groups at individually the sender and receiver to a different one by affixing some $A\upsilon$ virtual antennas without changing the physical array, based on the same assumptions as in [23]. The goal of AAVE is to progress the angular approximation resolution by adding simulated antennas to the existent of physical antennas at either sides of the Base Station (BS) and Mobile Terminal (MT).

2D DFT Approach: In [30], Millimeter Wave channel approximation technique relied on the Discrete Fourier Transform in two dimensional was offered. The approximation of the channel considerations is made by means of the iterative dissolution technique. In detail, the method after withdrawing the earlier estimated parameters in each iteration, samples are used in Discrete Fourier Transform (DFT) to examine the channel parameters for each path.

Switches Approach: In order to lessen price, intricacy and power intake, particularly at Mobile Terminals where these parameters are essential, in [31], the authors projected an innovative hybrid design for a Down Link Single User Millimeter Wave system by means of switches as a substitute for phase shifters.

Asymmetric Two Stage Approach: An unbalanced channel estimate procedure for a Down Link multi user Millimeter Wave system was suggested in [32]. As the name suggests it is a dual stage procedure namely, a comprehensive search phase, subsequently a CS approximation phase. During the former phase, the Base Station directs the training signals which points in all directions to the Mobile Terminals which in turn explore in comprehensive routine for the finest conjoining vectors to discover the Angle of Arrival. In the latter phase, CS approximation is done.

The Spatial Covariance Method: In [33], channel approximation algorithm scentered on assessing the three-dimensional covariance for a Time Division Duplexing Up Link Single User Millimeter Wave channel were offered. The researchers planned to evaluate the channel covariance straight without guesstimating the channel clearly whereas in two stage approach, first channel need to be approximated and followed by that finding out channel covariance need to be done.

Over Complete Dictionary Method: In [34] proposes an additional channel approximation approach for a fully-connected Up Link Single User mmWave system The number of Radio Frequency chains is significantly less than the number of antennas, and as a result, the signal at the receiver stage does not contain all of the Channel State Information, which the authors referred to as the measurement deficient that causes the channel approximation difficulty.

5 Conclusion

An overview of current channel estimate approaches for massive MIMO communication systems using mmWave is provided in this survey. First, the system architecture and channel characteristics that must be considered were explained. After that, we examined the performance of various channel estimate approaches in order to have a thorough understanding of each. This technique was used on millimeter Wave system architectures, specifically the hybrid design.

References

1. Agiwal, M., Roy, A., Saxena, N.: Next generation 5G wireless networks: a comprehensive survey. IEEE Commun. Surv. Tutor. **18**, 1617–1655 (2016)
2. Niu, Y., Li, Y., Jin, D., Su, L., Vasalakos, A.V.: A survey of millimeter wave communications (mmWave) for 5G: opportunities and challenges. Wirel. Netw. **1**, 2657–2676 (2015)
3. Xiao, M., et al.: Millimeter wave communications for future mobile networks. IEEE J. Sel. Areas Commun. **35**, 1909–1935 (2015)
4. Busari, S.A., Huq, M.S., Mumtaz, S., Dai, L., Rodriguez, J.: Millimeter-wave massive MIMO communication for future wireless systems: a survey. IEEE Commun. Surv. Tutor. **20**, 836–869 (2017)

5. Andrews, J.G., Bai, T., Kulkarni, M.N., Alkhateeb, A., Gupta, K., Heath, R.W.: Modeling and analyzing millimeter wave cellular systems. IEEE Trans. Commun. **65**, 403–430 (2016)
6. Hemadeh, I.A., Satyanarayana, K., El-Hajjar, M., Hanzo, L.: Millimeter-wave communications: physical channel models, design considerations, antenna constructions, and link-budget. IEEE Commun. Surv. Tutor. **20**, 870–913 (2017)
7. Wang, X., et al.: Millimeter wave communication: a comprehensive survey. IEEE Commun. Surv. Tutor. **20**, 1616–1653 (2018)
8. Rappaport, T.S., Heath Jr, R.W., Daniels, R.C., Murdock, J.N.: Millimeter wave wireless communications. Pearson Education, Boston (2015)
9. Szczypiorski, K.: HICCUPS: Hidden communication system for corrupted networks. In: International Multi-Conference on Advanced Computer Systems, pp. 31–40 (2003)
10. Ramesh Chandra, K., Borugadda, S.: An effective combination of terahertz band NOMA and MIMO system for power efficiency enhancement. Int. J. Commun. Syst. **35**(11), 83–89 (2022)
11. Zhu, M., Zhang, L., Wang, J., Cheng, L., Liu, C., Chang, G.K.: Radio-over-fiber access architecture for integrated broadband wireless services. J. Lightw. Technol. **31**, 3614–3620 (2013)
12. Rappaport, T.S., et al.: Wireless communications and applications above 100 GHz: Opportunities and challenges for 6G and beyond. IEEE Access **7**, 78729–78757 (2019)
13. Marcus, M., Pattan, B.: Millimeter wave propagation: spectrum management implications. IEEE Microw. Mag. **6**, 54–62 (2005)
14. Zhang, X., Andrews, J.G.: Downlink cellular network analysis with multi-slope path loss models. IEEE Trans. Commun. **63**, 1881–1894 (2015)
15. Sun, S., MacCartney, G.R., Rappaport, T.S. : Millimeter-wave distance-dependent large-scale propagation measurements and path loss models for outdoor and indoor 5G systems. In: IEEE 10th European Conference on Antennas and Propagation (EuCAP), pp. 1–5 (2016)
16. Banday, Y., Mohammad Rather, G., Begh, G.R.: Effect of atmospheric absorption on millimetre wave frequencies for 5G cellular networks. IET Commun. **13**, 265–270 (2019)
17. Pi, Z., Khan, F.: An introduction to millimeter-wave mobile broadband systems. IEEE Commun. Mag. **49**, 101–107 (2018)
18. Bjornson, E., Van der Perre, L., Buzzi, S., Larsson, E.G.: Massive MIMO in sub-6 GHz and mmWave: physical, practical, and use-case differences. IEEE Wirel. Commun. **26**, 100–108 (2019)
19. Doan, C.H., Emami, S., Sobel, D.A., Niknejad, A.M., Broadersen, R.W.: Design considerations for 60 GHz CMOS radios. IEEE Commun. Mag. **4**, 132–140 (2004)
20. Zhao, L.: Millimeter wave systems for wireless cellular communications. arXiv preprint arXiv: 1811. 1260 (2018)
21. Cui, S., Goldsmith, A.J., Bahai, A.: Energy-efficiency of MIMO and cooperative MIMO techniques in sensor networks. IEEE J. Sel. Areas Commun. **22**, 1089–1098 (2004)
22. Han, S., Chih-Lin, I., Xu, Z., Rowell, C.: Large-scale antenna systems with hybrid analog and digital beamforming for millimeter wave 5G. IEEE Commun. Mag. **53**, 186–194 (2015)
23. Alkhateeb, A., El Ayach, O., Leus, G., Heath, R.W.: Channel estimation and hybrid precoding for millimeter wave cellular systems. IEEE J. Sel. Top. Sign. Process. **8**, 831–846 (2014)
24. He, J., Kim, T., Ghauch, H., Liu, K., Wang, G., Millimeter wave MIMO channel tracking systems. In: IEEE Globecom Workshops (GC Wkshps), pp. 416–421 (2014)
25. Lee, J., Gil, G.T., Lee, Y.H.: Exploiting spatial sparsity for estimating channels of hybrid MIMO systems in millimeter wave communications. In: IEEE global Communications Conference, pp. 3326–3331 (2014)
26. Schniter, P., Sayeed, A.: Channel estimation and precoder design for millimeter-wave communications. In: 48th Asilomar Conference on Signals, Systems and Computers, pp. 273–277 (2014)

27. Payami, S., Shariat, M., Ghoraishi, M., Dianati, M.: Effective RF codebook design and channel estimation for millimeter wave communication systems. In: IEEE International Conference on Communication Workshop (ICCW), pp. 1226–1231 (2015)
28. Peng, Y., Li, Y., Wang, P.: An enhanced channel estimation method for millimeter wave systems with massive antenna arrays. IEEE Commun. Lett. **19**, 1592–1595 (2015)
29. Kokshook, M., Wang, P., Li, Y., Vucetic, B.: Fast channel estimation for millimetre wave wireless systems using overlapped beam patterns. In: IEEE International Conference on Communications (ICC), pp. 1304–1309 (2015)
30. Montagner, S., Benvenuto, N., Baracca, P.: Channel estimation using a 2D DFT for millimeter-wave systems. In: IEEE 81st Vehicular Technology Conference (VTC Spring), pp. 1–5 (2015)
31. Mendez-Rial, R., Rusu, C., Alkhateeb, A., Gonzalez-Prelcic, N., Heath, R.W. : Channel estimation and hybrid combining for mmWave: phase shifters or switches? In: Information Theory and Applications Workshop (ITA), pp. 90–97 (2015)
32. Noh, J., Kim, T., Seol, J.Y., Lee, C.: Zero-forcing based hybrid beamforming for multi-user millimeter wave systems. IET Commun. **10**, 2670–2677 (2016)
33. Park, S., Heath, R.W.: Spatial channel covariance estimation for mmWave hybrid MIMO architecture. In: 50th Asilomar Conference on Signals, Systems and Computers, pp. 1424–1428 (2016)
34. Guo, Z., Wang, X., Heng, W.: Millimeter-wave channel estimation based on 2-D beamspace MUSIC method. IEEE Trans. Wirel. Commun. **16**, 5384–5394 (2017)
35. Paul, P., Mowla, M.: A novel beamspace channel estimation technique for millimeter wave massive MIMO systems. In: International Conference on Electrical, Computer & Telecommunication Engineering (ICECTE), pp. 185–188 (2019)
36. Robaei, M., Akl, R., Chataut, R., Dey, U.K.: Adaptive millimeter-wave channel estimation and tracking. In: International Conference on Advanced Communication Technology (ICACT), pp. 23–28 (2022)

Comparison of Acoustic Channel Characteristics in Shallow and Deep-Sea Water

Ch. Venkateswara Rao[1](✉), M. Ravi Sankar[2], T. Nalini Prasad[3], R. Devi[3], M. Sri Uma Suseela[1], V. Praveena[1], and Y. Srinivas[1]

[1] Department of ECE, Vishnu Institute of Technology, Bhimavaram 534202, India
venkateswararao.c@vishnu.edu.in
[2] Department of ECE, Sasi Institute of Technology and Engineering, Tadepalligudem, India
[3] Department of ECE, SRKR Engineering College, Bhimavaram 534202, India

Abstract. The vital technology in the marine industry, underwater acoustic communication (UWAC), which has a crucial subsidiary role in undersea surveillance and military maneuvers. Now a days, the utmost prevalent and enduring system for undersea exploration is UWAC due to diminutive attenuation (signal reduction) of sound. However, the sound speed has abnormal variations with respect to the ocean columns. Usually, ocean water columns have been classified into two types based on the depth; such as shallow (0–100 m) and Deep-sea water (100-few thousands of kilometers). The physical (temperature, salinity) and chemical characteristics (pH, dissolved oxygen, nutrient salts) of these water divisions has abnormal variations with respect to the topographical regions and water depth. In addition, numerous factors such as; wave speed; multipath, interference, doppler spread have shown a great influence on sound propagation in underwater. Hence, a proficient channel modelling is required to analyze the effect of aforementioned parameters in order to achieve reliable communication in underwater environment. Therefore, in this work an acoustic channel model has been proposed for analyzing the sound speed variations and transmission losses in shallow and Deep-sea water scenarios. The effect of temperature, salinity on sound speed with respect to water depth has been estimated. Finally, the simulations outcomes of both shallow and Deep-sea water have been compared in terms of attenuation, sound speed, and transmission losses.

Keywords: Absorption · Attenuation · Acoustic Channel · Deepsea · Sound Speed · Temperature · Transmission Loss · Salinity

1 Introduction

Since they serve as enabling technology for a number of applications, underwater acoustic sensor networks (UASN) are gaining importance in the scientific community. The purpose of the underwater acoustic sensor networks (UASN), which are made up of submerged sensors, is to collect data on the portions of rivers and oceans that have not yet been extensively studied. A variable number of vehicles, anchored sensors, and

N. Gupta et al. (Eds.): IC4S 2022, LNICST 472, pp. 256–266, 2023.
https://doi.org/10.1007/978-3-031-28975-0_22

floating sensors are present in these networks, which are scattered throughout the study area [1]. These nodes create a channel of communication that includes one or more hops. Underwater devices can communicate with one another through optical, radio, electromagnetic, and acoustic waves [2].

Researchers like acoustic communication because it has a wide range and can convey digital data across an underwater channel. Significant research challenges are introduced by the network properties of the UASN (node mobility, transmission range, power), as well as the properties of the underwater medium (temperature, pressure, salinity, and pH), such as constrained bandwidth, multipath fading, constrained battery, and constrained data capacity [3]. Temperature, salinity, and sound speed with respect to depth are the key variables that affect an underwater habitat [4]. The undersea environment is also unpredictable because of a number of factors, including wave height, turbid currents, water pressure, water chemical compositions, and wave speed [5]. To create a trustworthy network, the suggested channel model must be able to detect changes in the properties of the underwater medium and network parameters. The operating frequency and sound speed of an acoustic channel model in an underwater environment typically affect how well it performs [6].

Underwater conditions can be influenced by temperature, salinity, depth, and pH [7, 8]. With respect to water depth, the sound speed varies as temperature and salinity change. The salinity of seawater varies with both water depth and location [9]. Salinity has been calculated from ocean dissolved salt concentrations and measured in parts per thousand (ppt). The establishment of links between the sensor nodes is impacted by these erratic changes in sound speed (based on temperature, salinity, water depth, and pH) and absorption (dependent on acoustic frequency, transmission distance). The creation of effective communication links between the sensor nodes is another factor which is influenced by change in sound speed [10, 11]. In order to ensure dependable communication between the sensor nodes in a deep underwater network, an efficient acoustic channel model that considers the impacts of sound speed, absorption losses, and transmission losses must be incorporated. Hence, this work is focused on developing a channel model which illustrates the comprehensive analysis of aforementioned channel characteristics in both the water divisions such as; shallow and Deep-sea scenarios.

2 Literature Review

Acoustic signal transmission through a water body as the route of propagation is significantly impacted by the marine environment. A few of the difficulties the acoustic channel faces are the Doppler shift, strong multi-path propagation, high attenuation, confined bandwidth, severe fading, extended delay spread, rapid time variation of the channel, route loss, and noise [12]. Therefore, it is crucial to conduct study and have a complete understanding of how the underwater environment influences the communication signal in both the regions (shallow and deep water) of the marine environment in order to construct and create reliable underwater communication systems. Although terrestrial networks have received a lot of attention from research teams, it is difficult to establish similar networks underwater due to the high absorption of electromagnetic radiation there. Additionally, there are several network constraints that separate terrestrial networks from underwater networks [13–24], including propagation delay, topology, power requirements, mobility, network life, data rate, etc.

Before comparing the issues and effects of employing different communication carriers, the authors in [25] supplied and explored the underlying physics of basic wave propagations using the fundamental physics ideas (acoustic, EM, and optical). The effects of propagation properties, such as sound speed, channel delay, absorption, scattering, multipath, waveguide effects, and ambient noise, have been studied by the authors in [26]. The dependence of the channel capacity on depth and temperature was examined by the authors in [27] using enhanced propagation loss and ambient noise models. A mathematical model [28] that allows the conversion of atmospheric pressure to depth and depth to atmospheric pressure has been presented in order to reduce errors in the sound speed in oceans and seas. An experimental setup that demonstrates how salinity, temperature, and pressure affect the physical characteristics of the deployed environment and sound speed fluctuations is detailed in [29]. A technique for raising localization accuracy [30] in water involves calculating the sound speed at a particular location over time. An acoustic channel model has been created to simulate networks in aquatic environments [31].

An underwater acoustic channel path loss has been measured in real-time [32]. Recent advances in deep learning and artificial intelligence have been used to predict the characteristics of underwater acoustic channels in order to increase accuracy and throughput. A deep learning-based framework for underwater channel modelling has been presented [33] to improve the accuracy of channel models. The channel model's primary statistical properties have been outlined and discussed in [34]. The literature has provided insight into how variations in the underwater medium's characteristics impact underwater sound propagation. Therefore, it is important to address and take into consideration the effects of temperature, salinity, absorption, and transmission losses brought on by changes in the sound speed. These variations in sound speed brought on by the unpredictability of the undersea environment change how links grow in the network. Since transmission losses and absorption must be considered, the suggested study objective is to analyze and compare the acoustic channel parameters in shallow and Deep-sea scenarios.

3 Methodology

The influence of underwater physical parameters such as temperature and salinity on sound speed has been estimated using designed acoustic channel model for shallow and Deepsea water scenarios. The channel model evaluates the attenuation which is attained because of chemical compositions of the sea water. The transmission losses for both shallow and Deepsea water scenarios have been evaluated with respect to the frequency and distance. The performance measures of proposed channel model have been compared in terms of sound speed, attenuation, and transmission losses for both shallow and Deepsea water scenarios.

3.1 Sound Velocity

One of the key distinctions between acoustic waves and EM transmission is the extraordinarily slow speed at which sound travels through water. Pressure, salinity, and temperature are a few factors that affect the undersea sound speed. Normal ocean surface sound propagation occurs at a speed of around 1520 m/s, which is four orders of magnitudes quicker than airborne sound propagation but five orders of magnitude slower than light propagation. Environmental changes have a direct impact on the sound speed of water (temperature, salinity, and depth). For computing sound velocity as a function of temperature, salinity, and depth, Mackenzie presents an empirical formula [35] which is epitomized using (1).

$$c(T, S, z) = a_1 + a_2T + a_3T^2 + a_4T^3 + a_5(S - 35) + a_6z + a_7z^2 + a_8T(S - 35)a_9Tz^3 \tag{1}$$

3.2 Propagation Loss of Sound

Undersea acoustics attenuate due to spreading, dispersion, and absorption. A measure of signal degradation brought on by the geometrical spreading effect that happens as a sound wave travels away from its source is called spreading loss. In underwater acoustics, there are two basic kinds of spreading mechanisms: cylindrical and spherical. The loss is expressed using (2) for cylindrical spreading and (3) for spherical spreading, respectively. L_{CS}, L_{SS}, and R_t stand for transmission range, loss due to cylindrical spreading, and loss due to spherical spreading, respectively [36].

$$L_{CS} = 10\log(R_t) \tag{2}$$

$$L_{SS} = 20\log(R_t) \tag{3}$$

3.3 Absorption Coefficient

An empirical formula [36] for the absorption coefficient has been used for the frequency range of 100 Hz to 1 MHz which is represented using (4) and it is a function of frequency, pressure (depth), and temperature.

$$\alpha = \frac{A_1P_1f_1f^2}{f^2 + f_1^2} + \frac{A_2P_2f_2f^2}{f^2 + f_2^2} + A_3P_3f^2 \tag{4}$$

3.4 Absorption Loss

The range-dependent absorption loss, which is computed using Eq. (5), indicates the energy loss of sound as a result of the conversion of energy into heat due to the chemical properties of viscous friction and ionic relaxation in the ocean.

$$L_{ab} = (\alpha \times R_t) \times 10^{-3} \tag{5}$$

3.5 Transmission Loss

Transmission loss [37], which is defined as the cumulative loss of acoustic strength caused by an acoustic pressure wave as it moves away from its source which is represented by using (6) & (7) for both shallow and Deep-sea water scenarios respectively.

$$TL_{shallow} = L_{CS} + L_{ab} \quad (6)$$

$$TL_{Deepsea} = L_{SS} + L_{ab} \quad (7)$$

4 Implementation Parameters

The parameters considered for analyzing the transmission loss, absorption loss, spreading loss and sound speed has been listed in Table 1.

Table 1. Implementation Parameters

Parameter	Range
Shallow water Depth	0–100 m
Deep water Depth	100–8000 m
Shallow Water Temperature	30–24°C
Deep Water Temperature	24°–4°C
Shallow Water Salinity	30–35 ppt
Deep Water Salinity	35 ppt
Frequency	100 Hz–100 kHz
pH	7.8
R_t	100 m

5 Simulation Results

The primary method of wireless data transfer in a marine environment is the acoustic medium. The most fundamental factor affecting achievable data rates, as well as network factors like latency and quality of service, is the speed of sound in an acoustic channel. Due to irregular variations in temperature, salinity, depth, and pH with respect to the season, time, and location of the ocean, the sound speed has unpredictable variations underwater. However, these irregularities have different profiles in the two divisions of the ocean such as shallow and deep water. In, shallow water the effect of temperature and salinity are predominant factor for abrupt change in sound speed, where as in deep water the effect of temperature and salinity on sound speed is insignificant due to constant values of temperature (4 °C) and salinity (35 ppt) in deep water. In deep water, the effect of pressure (depth) is significant on sound speed. The sound speed profile exhibits abnormal variations in shallow water because of drastic change in temperature gradients of the water column across the water depth (See Fig. 1). Whereas, the temperature is almost constant (4 °C) at the deep water scenarios, where the sound speed profile has minimal variations. It is clearly depicted in Fig. 1, that the sound speed is varying with temperature and depth. When the depth is extended to 100 m, the sound speed increases to 1542 m/s (see Fig. 1) from the initial value of 1539 m/s at a certain temperature and depth ($T = 27^{\circ}$C, $D = 100$ m).

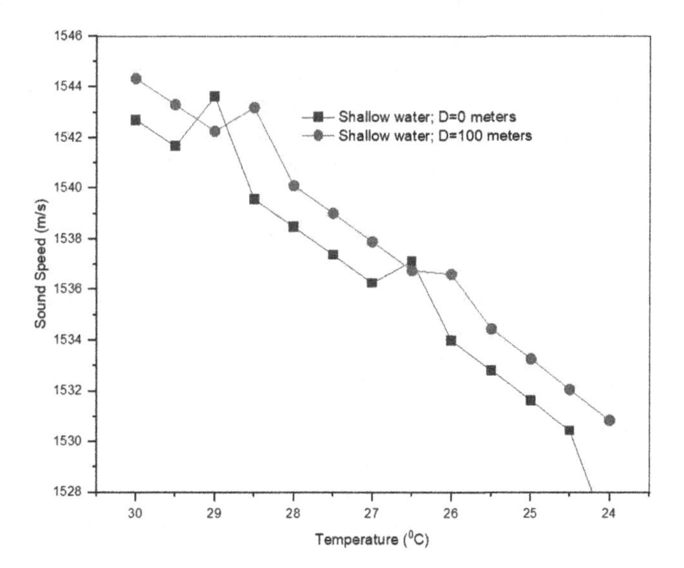

Fig. 1. Effect of temperature on sound speed in shallow water.

In the case of deep water, When the depth is extended to 8000 m and the temperature is lowered to 4°C, the sound speed increases to 1450 m/s (see Fig. 2) from the initial value of 1530 m/s at a certain temperature and depth ($T = 4^{\circ}$C, $D = 8000$ m).

Similarly, salinity of the ocean water increases along the depth, which also influences the sound speed in deep water. This is clearly depicted in Fig. 3, that the sound speed increases with increase in depth as well as salinity in shallow water. At a particular salinity (S = 33 ppt), the sound speed attained different profiles (varying from 1540 m/s to 1550 m/s) along the depth (see Fig. 3). Whereas, in deep water the salinity value is almost constant (s = 35 ppt), the sound speed has liner in nature, which is increases with depth (see Fig. 4).

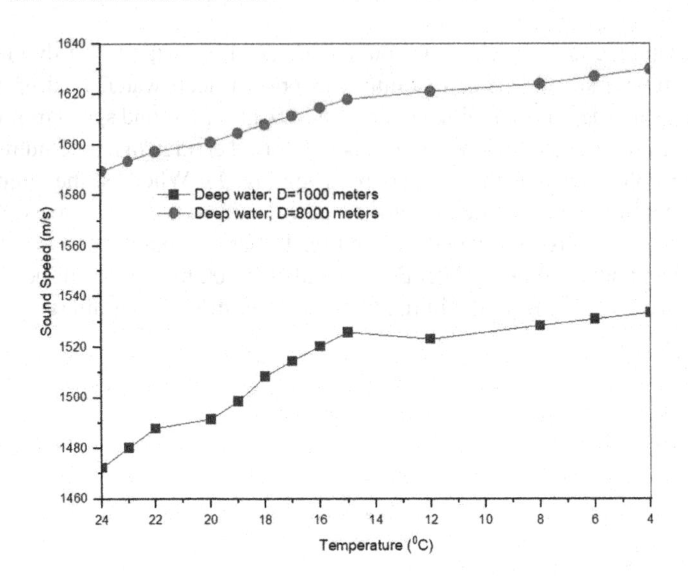

Fig. 2. Effect of temperature on sound speed in Deep water.

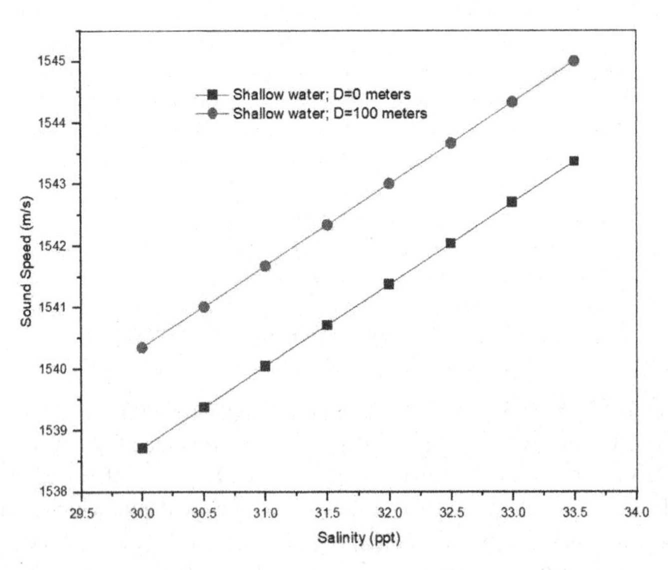

Fig. 3. Effect of salinity on sound speed in shallow water.

Absorption, which results from the transformation of acoustic energy into heat, is the principal cause of attenuation. As the distance and frequency rise, the attenuation grows. It is also obvious that the orders of magnitude vary widely and that attenuation rises sharply with frequency. For frequencies of 1 kHz and below, attenuation is less than a few hundredths of a dB/km; hence, it is not a limiting factor. Approximately 1 dB/km

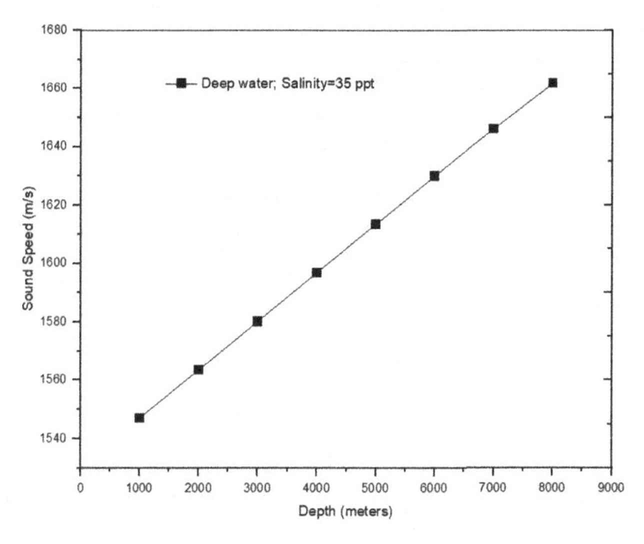

Fig. 4. Effect of salinity on sound speed in Deep water.

of attenuation occurs at 10 kHz, restricting ranges of more than a few tens of kilometres (see Fig. 5).

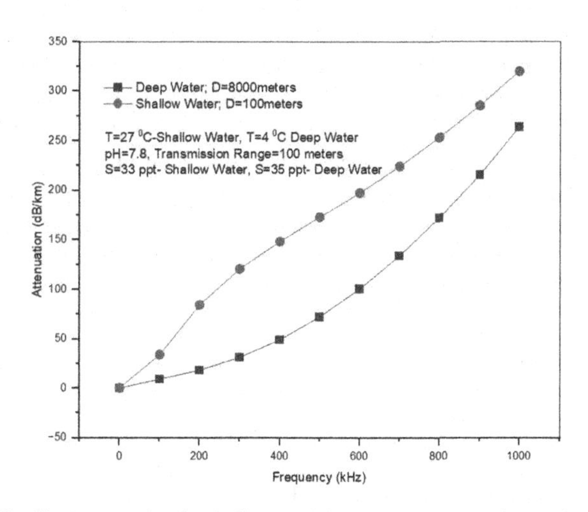

Fig. 5. Attenuation in shallow and deep water due to absorption.

The transmission losses are frequency and range dependent. According to Fig. 6, transmission losses increase as frequency increases, while transmission losses decrease as depth increases. The transmission losses are higher in shallow water when compared to deep water because of highly abnormal environment conditions in shallow water (see Fig. 6).

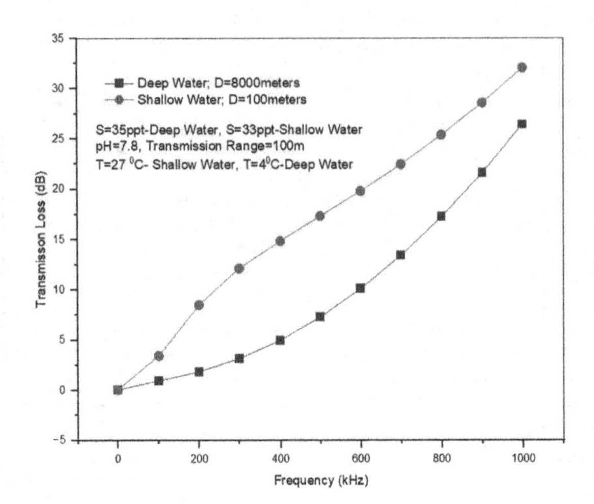

Fig. 6. Transmission losses with respect to frequency in shallow and deep water.

6 Conclusion

In order to examine the impact of underwater medium characteristics like temperature and salinity on sound speed, this study suggests an acoustic channel model for both ocean divisions. The suggested channel model analyses the influence of salinity and temperature at a given pH by adjusting different depths in shallow and Deep-sea scenarios. The effect of absorption resulting from various water chemical compositions with regard to the frequency of the acoustic signal is also investigated by the proposed channel model. Finally, with a fixed transmission range and various depths in shallow and deep water, the transmission losses have been evaluated. According to the simulation results, transmission and absorption losses depend on frequency. The frequency has grown proportionately to these two losses. In deep water, the sound speed rises as temperature and salinity gradually decrease with depth. Whereas, in shallow water the sound speed has been gradually decreased with respect to depth.

References

1. Sozer, E.M., Stojanovic, M., Proakis, J.G.: Underwater acoustic networks. IEEE J. Ocean Eng. **25**(1), 72–83 (2000)
2. Akyildiz, I.F., Pompili, D., Melodia, T.: Underwater acoustic sensor networks: research challenges. Ad Hoc Netw. **3**(3), 257–279 (2005)
3. Barbeau, M., Garcia-Alfaro, J., Kranakis, E., Porretta, S.: The sound of communication in underwater acoustic sensor networks. In: Zhou, Y., Kunz, T. (eds.) Ad Hoc Networks. LNICSSITE, vol. 223, pp. 13–23. Springer, Cham (2018). https://doi.org/10.1007/978-3-319-74439-1_2
4. Akyildiz, I.F., Pompili, D., Melodia, T.: Challenges for efficient communication in underwater acoustic sensor networks. ACM SIGBED Rev. Spec. Issue Embed. Sensor Netw. Wirel. Comput. **1**(2), 3–8 (2004)

5. Stojanovic, M., Preisig, J.: Underwater acoustic communication channels: propagation models and statistical characterization. IEEE Commun. Mag. **47**(1), 84–89 (2009)
6. Jindal, H., Saxena, S., Singh, S.: Challenges and issues in underwater acoustics sensor networks: a review. In: 2014 International Conference on Parallel, Distributed and Grid Computing Solan, pp. 251 255 (2014)
7. Ismail, N.S.N., Hussein, L.A., Ariffin, S.H.: Analyzing the performance of acoustic channel in underwater wireless sensor network. In: Asia International Conference on Modelling and Simulation, pp. 550–555 (2010)
8. Wanga, X., Khazaiec, S., Chena, X.: Linear approximation of underwater sound speed profile: Precision analysis in direct and inverse problems. Appl. Acoust. **140**, 63–73 (2018)
9. Ali, M.M., Sarika, J., Ramachandran, R.: Effect of temperature and salinity on sound speed in the central Arabian sea. Open Ocean Eng. J. **4**, 71–76 (2011)
10. Kumar, S., Prince, S., Aravind, J.V., Kumar, G.S.: Analysis on the effect of salinity in underwater wireless optical communication. Mar. Georesour. Geotechnol. **38**(3), 291–301 (2020)
11. Hovem, J.: Underwater acoustics: propagation, devices and systems. J. Electroceram. **19**, 339–347 (2007)
12. Chitre, M., Shahabodeen, S., Stojanovic, M.: Underwater acoustic communications and networking: recent advances and future challenges. Mar. Technol. Soc. J. **42**, 103–116 (2008)
13. Chaturvedi, S.K., Padmavathy, N.: The influence of scenario metrics on network reliability of mobile ad hoc network. Int. J. Performability Eng. **9**(1), 61–74 (2013)
14. Kumar, B.V.S., Padmavathy, N.: A hybrid link reliability model for estimating path reliability of mobile ad hoc network. Procedia Comput. Sci. **171**, 2177–2185 (2020)
15. Cook, J.L., Ramirez-Marquez, J.E.: Two-terminal reliability analyses for a mobile ad hoc wireless network. Reliab. Eng. Syst. Saf. **92**(6), 821–829 (2007)
16. Padmavathy, N., Chaturvedi, S.K.: A systematic approach for evaluating the reliability metrics of MANET in shadow fading environment using monte Carlo simulation. Int. J. Performability Eng. **12**, 265–282 (2016)
17. Padmavathy, N., Chaturvedi, S.K.: Reliability evaluation of capacitated mobile ad hoc network using log-normal shadowing propagation model. Int. J. Reliab. Saf. **9**(1), 70–89 (2015)
18. Venkata Sai, B., Padmavathy, N.: A systematic approach for analyzing hop count and path reliability of mobile ad hoc networks. In: International Conference on Advances in Computing, Communications and Informatics, pp. 155–160 (2017)
19. Padmavathy, N., Anusha, K.: Dynamic reliability evaluation framework for mobile ad-hoc network with non-stationary node distribution. In: Communication and Computing Systems. CRC Press, Taylor and Francis (2018)
20. Padmavathy, N., Teja, J R C., Chaturvedi, S K.: Performance evaluation of mobile ad hoc network using MonteCarlo simulation with failed nodes. In: 2ndInternational Conference on Electrical, Computer and Communication Technologies, pp. 1–6 (2017)
21. Cook, J.L., Ramirez-Marquez, J.E.: Reliability analysis of cluster-based ad-hoc networks. Reliab. Eng. Syst. Saf. **93**(10), 1512–1522 (2008)
22. Padmavathy, N.: An efficient distance model for the estimation of the mobile ad hoc network reliability. In: Gunjan, V. K., Garcia Diaz, V., Cardona, M., Solanki, V. K., Sunitha, K. V. N. (eds.) ICICCT 2019, pp. 65–74. Springer, Singapore (2020). https://doi.org/10.1007/978-981-13-8461-5_8
23. Venkateswara Rao, C., Padmavathy, N.: Effect of link reliability and interference on two-terminal reliability of mobile ad hoc network. In: Verma, P., Charan, C., Fernando, X., Ganesan, S. (eds.) Advances in Data Computing, Communication and Security. LNDECT, vol. 106, pp. 555–565. Springer, Singapore (2022). https://doi.org/10.1007/978-981-16-8403-6_51

24. Rao, C V., Padmavathy, N., Chaturvedi, S K.: Reliability evaluation of mobile ad hoc networks: with and without interference. In: IEEE 7th International Advance Computing Conference, pp. 233–238 (2017)
25. Lanbo, L., Shengli, Z., Jun-Hong, C.: Prospects and problems of wireless communication for underwater sensor networks. Wirel. Commun. Mob. Comput. **8**, 977–994 (2008)
26. Preisig, J.: Acoustic propagation considerations for underwater acoustic communications network development. Mob. Comput. Commun. Rev. **11**(4), 2–10 (2006)
27. Sehgal, A., Tumar, I., Schonwalder, J.: "Variability of available capacity due to the effects of depth and temperature in the underwater acoustic communication channel. In: *OCEANS 2009-EUROPE*, Bremen, pp. 1–6 (2009)
28. Leroy, C.C., Parthiot, F.: Depth-pressure relationships in the oceans and seas. J. Acoust. Soc. Am. **103**(3), 1346–1352 (1998)
29. Yuwono, N P., Arifianto, D., Widjiati, E., Wirawan.: Underwater sound propagation characteristics at mini underwater test tank with varied salinity and temperature. In: 6th International Conference on Information Technology and Electrical Engineering (ICITEE), pp. 1–5 (2014)
30. Shi, H., Kruger, D., Nickerson, J V.: Incorporating environmental information into underwater acoustic sensor coverage estimation in estuaries. In: MILCOM 2007 - IEEE Military Communications Conference, pp. 1–7 (2007)
31. Morozs, N., Gorma, W., Henson, B.T., Shen, L., Mitchell, P.D., Zakharov, Y.V.: Channel modeling for underwater acoustic network simulation. IEEE Access **8**, 136151–136175 (2020)
32. Lee, H.K., Lee, B.M.: An underwater acoustic channel modeling for internet of things networks. Wireless Pers. Commun. **116**(3), 2697–2722 (2020). https://doi.org/10.1007/s11277-020-07817-x
33. Onasami, O., Adesina, D., Qian, L.: Underwater acoustic communication channel modeling using deep learning. In: 15th International Conference on Underwater Networks & Systems (WUWNet 2021), China (2021)
34. Zhu, X., Wang, C.X., Ma, R.: A 2D non-stationary channel model for underwater acoustic communication systems. In: IEEE 93rd Vehicular Technology Conference (VTC2021-Spring), pp. 1–6 (2021)
35. Mackenzie, K.V.: Nine-term equation for sound speed in the oceans. J. Acoust. Soc. Am. **70**(3), 807–812 (1981)
36. Etter, P.C.: Underwater Acoustic Modeling and Simulation, 3rd edn. Spon Press, New York (2003)
37. Padmavathy, N., Ch, V.R.: Reliability evaluation of underwater sensor network in shallow water based on propagation model. In: Journal of Physics: Conference Series, vol. 1921, p. 012018 (2021). 1–17

Author Index

N. Gupta et al. (Eds.): IC4S 2022, LNICST 472, pp. 267–268, 2023.
https://doi.org/10.1007/978-3-031-28975-0

Printed in the United States
by Baker & Taylor Publisher Services